INTRODUCTION TO ENVIRONMENTAL REMOTE SENSING

INTRODUCTION TO ENVIRONMENTAL REMOTE SENSING

Fourth Edition

E.C. Barrett

Senior Research Fellow and Formerly Director
Centre for Remote Sensing
School of Geographical Sciences
University of Bristol

and

L.F. Curtis OBE

Formerly Exmoor National Park Officer and
Honorary Senior Research Fellow
Department of Geography
University of Bristol

Stanley Thornes (Publishers) Ltd

Third edition published by Chapman & Hall in 1992
(ISBN 0 412 37170 7)

Fourth edition published in 1999 by:
Stanley Thornes (Publishers) Ltd
Ellenborough House
Wellington Street
Cheltenham
Glos.
GL50 1YW
United Kingdom

99 00 01 02 03 / 10 9 8 7 6 5 4 3 2 1

A catalogue record for this book is available from the British Library

ISBN 0 7487 4006 6

Typeset by WestKey Ltd, Falmouth, Cornwall
Printed in Croatia by Zrinski d.d., Čakovec

To the memory of Leonard F. Curtis, 1924–1991,
remote sensing pioneer, distinguished teacher, effective manager,
skilful diplomat, colleague and good friend.

CONTENTS

PART TWO REMOTE SENSING APPLICATIONS

x   *Contents*

PREFACE

In the Preface to the second edition of this volume, published in 1982, Dr Curtis and I referred to an 'explosive growth' of remote sensing of the environment, for such a description was fully merited. Sixteen years and two further editions later the same description is still appropriate, indeed all the more so, for environmental remote sensing continues to grow apace and, with a galaxy of new platforms beginning to pepper the skies, further rapid growth is certain – a fact ever more strongly attested by the commercial as well as the scientific worlds.

Whilst such impressive vigour proffers and promises much to those who, for any one of many possible reasons, are now interested in remote sensing, it also poses some difficult problems, not least for book authors and publishers. One such problem is how to select material for an introductory text that is still broad enough to fulfil its tutorial purposes yet compact enough to remain within the purchasing power of students: whole libraries of remote sensing volumes and journals are now springing up! A second problem is how best to update a text in this field within all the necessary but stringent sets of accompanying constraints, rather than to wipe the slate clean and start again. In the present case so much new material has clamoured for inclusion that I have often been on the point of giving up the revision task.

Fortunately such problems are offset by special opportunities. One of these is that, as any field of study becomes broader and deeper, so the need for self-contained books to describe it as a whole – from first principles, through to present uses, and on to future possibilities – becomes not less, but greater than before. Another is that entirely new books run the risk of painting pictures of scientific and/or technological trees without their roots – often failing to do justice to the all-important ways in which key philosophies, principles and practices have grown to what they are today. All of us who have been involved with the first three editions of the present title have been pleased and encouraged by the wide circulation of this book, and its acceptance as a standard text in its field. The preparation of this fourth edition has proved a fascinating task – challenging, yes, but also for several reasons much more difficult than originally expected.

Since the field of environmental remote sensing is now so extensive, this new edition must be seen as a *beginning* for students who can afford to purchase one book but not several; a *basis* for staff who have broad courses to design and deliver; and a *lifeline* for career professionals whose formal training did not include any or enough information on this youthful discipline to fit them for present-day decision making.

My solution to the challenge of how to simultaneously acknowledge earlier work in remote sensing whilst bringing the story up to date has been to retain text, illustrations and references where the pioneering work so described merits this even today, but to replace it, or to add to it, where principles are now differently understood, and/or systems or applications have significantly advanced. In such regards, it is especially noteworthy that the early work in most applications areas not only has a habit of identifying and addressing the central issues, but also often succeeds in taking the biggest single step towards achieving worthwhile results. In the Bibliography, this balance of 'old' and 'new' mostly entails papers from the pioneering past, and books from the more recent past and presents a pattern dictated largely by the very rapid rate of growth in the remote sensing literature from then until now.

In the Preface to the third edition I expressed the shock that I and many others felt when we heard of Len Curtis's sudden, unexpected and untimely death, at an advanced stage in the preparation of that book. This fourth edition, designed to update the third, has therefore been in one sense all my own work, and a lonely task at that. However, considerable sections of the text rightly survive from Len's own inputs. All too often today due credit is denied the earlier work and workers in every scientific and technological field. My personal belief is that modern texts do well to blend more recent results with the pioneering studies: the latter are the foundations for the former, and are often easier to assimilate on account of their relative directness and simplicity. Therefore, this book continues to benefit from the broad personal knowledge and experience of remote sensing of both its original authors, for both of us became actively involved in it soon after its name was coined in 1960; both of us have devoted lengthy spells of our respective scientific careers to remote sensing research and applications. It is my hope and belief that, not least by spreading credit over scientists of two whole generations, this new edition will continue to communicate the essence of remote sensing clearly and honestly to new classes of students – the professional remote sensing experts of the future – as well as to the growing ranks of professionals in other fields who wish or need to understand what it is about.

I must include in my notes of thanks to others my deep gratitude to Len's widow, Mrs Diana Curtis, for supporting and helping to expedite the contracting of this new edition. She continues in no small way to maintain the Curtis tradition of service to the community, particularly in the Exmoor region which Len himself loved and served so well.

I owe great debts of gratitude also to the many friends and colleagues around the world from whom Dr Curtis and I have learnt much about remote sensing, and with whom we have at different times discussed matters covered by this title. Where direct use has been made of the work of others, acknowledgement is given through captions and references; the latter have been augmented by other selected items which will help to widen the horizons of the reader.

Particular thanks are due to our fourth edition referees, and to our past and present friends and colleagues in the University of Bristol, the Exmoor National Park Office, the UK Meteorological Office, the National Rivers Authority, the European Association of Remote Sensing Laboratories, the European Space Agency, the US National Aeronautics and Space Administration, and the US National Oceanic and Atmospheric Administration, among others, for their support and encouragement over the years, and for allowing access to material used in this book. We trust that they, and all others whose work we have drawn from, will feel that their findings have been represented fairly – if much more briefly than either they or we would have liked. Mrs Shirley Sparks and my wife Gillian have wonderfully supported me on their word-processors, and in other practical ways, including just the right measures of sympathy, understanding and encouragement when these were so urgently needed. I am indebted to all these good folk, and more besides.

For the sake of Mrs Diana Curtis and her daughters Sarah and Ruth, and for the continuing memory of a great colleague and very good friend, I hope this book will go on being a successful, and therefore fitting, tribute to Len.

E.C. Barrett
Backwell
North Somerset

THE AUTHORS

DR ERIC C. BARRETT

Dr Eric Barrett is a Senior Research Fellow in the top-ranked (5*) School of Geographical Sciences in the University of Bristol, UK, having served as Founding Director both of its Remote Sensing Unit from 1983, and of its successor, the University Research Centre of Remote Sensing, from 1994 to 1996. He is author or editor of over 20 books or monographs on geography, climatology, hydrology or remote sensing, in addition to over 300 papers, articles and reports mainly in these fields. His interest in remote sensing was aroused first by a presentation in London in 1963 by the chief of the TIROS Project, which was then in its infancy. This prompted Dr Barrett's first book, entitled *Viewing Weather from Space*, published in 1967, and led to his PhD on 'The Contribution of Meteorological Satellites to Dynamic Climatology' in 1969.

Since then, Dr Barrett's interests and involvements in remote sensing have been wide and varied. He has undertaken many consultancies for commercial companies, national governments and international agencies, including FAO, UNDRO and WMO, and for 6 years was Rapporteur to UNESCO on remote sensing applications in hydrology and water management. He has served on numerous expert committees on remote sensing in the UK, Europe and the USA, and has played an active role in many pioneering projects, including the Landsat 2 Program (as one of only two UK Principal Investigators), the US AgRISTARS Project, NASA's WetNet Project (serving for 7 years as its Precipitation Working Group Chairman) and ESA's ERS-1 Programme. In 1982 he was awarded the Hugh Robert Mill Medal and Prize of the Royal Meteorological Society for the development of methods for rainfall monitoring by satellites, and also the DSc 'advanced doctorate' by the University of Bristol, for his 'sustained and distinguished contribution to geographic science'. Once appointed Director of the newly established Remote Sensing Unit in that university, he built this, and the more recent Centre for Remote Sensing, to become one of the foremost groups of its kind, internationally recognized for its work particularly in the area closest to his own heart: applied satellite hydrometeorology.

DR LEONARD F. CURTIS

Dr Leonard Curtis was Reader in Geography and Head of the Joint School of Botany and Geography in the University of Bristol before leaving in 1978 to become National Park Officer for Exmoor. He was invested by HRH Queen Elizabeth II with the Order of the British Empire on his retirement from the Exmoor National Park Office in 1988. His early experience of remote sensing began with wartime air photography in the Far East, during service with the RAF, first as Navigator, then as Squadron Leader by the time of demobilization. Commercial experience of applied photo-interpretation followed during a period of secondment from the Soil Survey of England and Wales to Huntings Surveys, involving work on irrigation development studies. After joining the Department of Geography in Bristol University as Lecturer in 1956 his interests in remote sensing broadened rapidly, and he served on a number of early applications committees, including those of the European Space Research Organization and its successor, the European Space Agency. He later served as consultant to many UK and European programmes and projects in remote sensing, and quite recently as Secretary-General and Vice Chairman of EARSeL (the European Association of Remote Sensing Laboratories), a body which he had helped to found in the late 1970s.

Dr Curtis authored or edited more than 10 books concerned with remote sensing applications, soil studies and land use inventories, and contributed greatly to the development of uses of remote sensing in environmental monitoring and management in protected landscapes. Very sadly, he died suddenly in April 1991 at the age of 67. The Earth Observation Quarterly, published by ESA, said of him in its subsequent obituary: 'Len Curtis, without fuss or bother, carved a niche for himself amongst the pioneers of European remote sensing'. This book is one of the many monuments to his life and work.

ABBREVIATIONS AND ACRONYMS

ACT   Automatic Collection and Telemetry
ADEOS   Advanced Earth Observation Satellite
AEM-A   Applications Explorer Mission-A
AgRISTARS   Agriculture and Resources Inventory Surveys Through Aerospace Remote Sensing
AID   Agency for International Development (USA)
AIS   Airborne Imaging Spectrometer
AMI   Active Microwave Instrument
AMSR   Advanced Mechanically Scanning Radiometer
AMSU   Advanced Microwave Sounding Unit
ARTEMIS   Africa Real-Time Environmental Monitoring and Information System
API   Air Photo Interpretation
APT   Automatic Picture Transmission
ATS   Applications Technology Satellite
ATSR   Along-Track Scanning Radiometer
AVHRR   Advanced Very High Resolution Radiometer
AVIRIS   Airborne Visible/Infrared Imaging Spectrometer

BDR   Bidirectional Reflectance
BF   Banding Feature (of hurricane)
BNSC   British National Space Centre
BRDF   Bidirectional Reflectance Distribution Function
BRF   Bidirectional Reflectance Factor
BSU   Basic Sounding Unit

CAD   Computer Aided Design
CAT   Clear Air Turbulence
CCT   Computer-Compatible Tape
CCEW   Countryside Commission for England and Wales
CDO   Central Dense Overcase (of hurricane)
CEC   Commission of the European Communities
CEOS   Committee on Earth Observation Satellites
CFC   Chlorofluorocarbon
CNES   Central Nationale d'Etudes Spatiales (France)
CORINE   Coordination, Information, Environment (project of the CEC)
COSPAR   Committee for Space Research (UN)
CPU   Central Processing Unit
CRT   Cathode Ray Tube

CZCS	Coastal Zone Colour Scanner
DBMS	Database Management System
DC	Direct Current
DCP	Data Collection Platform
DCS	Data Collection System
DEM	Digital Elevation Model
DLCC	Desert Locust Control Commission
DMSP	Defense Meteorological Satellite Program
DoE	Department of the Environment (UK)
DPA	Drainage Pattern Analysis
DTM	Digital Terrain Model
DVP	Digital Video Plotter
EARSeL	European Association of Remote Sensing Laboratories
EBR	Electronic Beam Recorder
ecu	European Currency Unit
EDC	EROS Data Center (USA)
EDIS	Environmental Data and Informative Service (USA)
E (E) C	European (Economic) Community
EMI	Electro-Magnetic Industries, plc
EOS	Earth Observation System
EOSAT	Earth Observation Satellite
EPOP	European Polar Operational Platform
ERB	Earth Radiation Budget
ERBE	Earth Radiation Budget Experiment
ERBS	Earth Radiation Budget Satellite
ERDAS	Earth Resources Data Analysis System
EROS	Earth Resources Observation System
ERS	Environmental Remote Sensing; European Resources Satellite
ERS-1	Earth Resources Satellite
ESA	European Space Agency
ESMR	Electronically-Scanning Microwave Radiometer
ESOC	European Space Operations Centre (Darmstadt, Germany)
ESSA	Environmental Sciences Services Administration
ESTAR	Electronically-Scanning Thinned Array Radiometer
ET	Evapotranspiration
ETM	Enhanced Thematic Mapper
EUMETSAT	European Meteorological Satellite
EURASEP	European Association of Scientists in Environmental Pollution
FAO	Food & Agriculture Organisation of the United Nations
FAS	Foreign Agricultural Service
FFT	Fast Fourier Transform
FGGE	First GARP Global Experiment
FIFE	First ISLSCP Field Experiment
FLIR	Forward-Looking Infrared Radiometer

FOV	Field of View
FRONTIERS	Forecasting Rain, Optimized using New Techniques of Interactively Enhanced Radar & Satellite data
GALE	Genesis of Atlantic Lows Experiment
GAP	Geographic Applications Program (USGS)
GARP	Global Atmospheric Research Program
GATE	GARP Atlantic Tropical Experiment
GDD	Growing Degree Day
GE	Ground Enumeration
GEMS	Global Environment Monitoring System
GEWEX	Global Energy and Water Experiment
GGEM	Gravity Gradiometer Explorer Mission
GHz	Giga-Hertz
GIS	Geographic Information System
GLAI	Green Leaf Area Indices
GLU	Grazing Livestock Unit
GMS	Geostationary Meteorological Satellite
GMT	Greenwich Mean Time
GOES	Geostationary Operational Environmental Satellite
GOES-IO	Global Operational Environmental Satellite, Indian Ocean
GOME	Global Ozone Monitoring Equipment
GOOS	Global Ocean Observing System
GOSSTCOMP	Global Operational Sea Surface Temperature Computation
GPS	Global Positioning System
GRE	Ground Resolution Element
GSFC	Goddard Space Flight Center (Greenbelt, MD)
HCMM	Heat Capacity Mapping Mission
HCMR	Heat Capacity Mapping Radiometer
HIPLEX	High Plains Experiment
HIRIS	High Resolution Imaging Spectrometer
HRIS	High-Resolution Infrared Sounder
HRPT	High Resolution Picture Transmission
HRV	High Resolution Visible
IBP	International Biological Programme
IC	Image Count
ICSU	International Council of Scientific Unions
IFFA	Interactive Flash-Flood Analysis
IFOV	Instantaneous Field of View
IGBP	International Geosphere–Biosphere Programme
IMC	Image-Motion Compensation
IPIPS	Infrastructure Planetary Image Processing System
IR	Infrared
IRIS	InfraRed Interferometer Spectrometer
IRLS	Infrared Linescanner

IRS	Indian Remote Sensing Satellite
IR/VIS	Infared/Visible
I²S	International Imaging Systems
ISLSCP	International Satellite Land Surface Climatology Project
ISPRS	International Society of Photogrammetry and Remote Sensing
ITCB	Intertropical Cloud Band
ITCZ	Intertropical Convergence Zone
ITOS	Improved TIROS Operational Satellite
JERS	Japanese Earth Resources Satellite
JPL	Jet Propulsion Laboratory (USA)
JPOP	Japanese Polar Operational Platform
JRC	Joint Research Centre (EEC, Ispra)
LAC	Large Area Coverage (NOAA data)
LACIE	Large Area Crop Inventory Experiment
LAGEOS	Laser Geodetic Earth-Orbiting Satellite
LAI	Leaf Area Index
LARS	Laboratory for Agricultural Remote Sensing (Purdue University, US)
LAWS	Laser Atmospheric Wind Sounder
LBA	Large-scale Biosphere/Atmosphere Programme, Amazonia
LFC	Large Format Camera
LFMR	Low Frequency Microwave Radiometer
LIMS	Limb Infrared Monitoring of the Stratosphere
LISS	Linear Self-Scanning
LRR	Laser Retro-Reflector
LST	Local Solar Time
MARS	Monitoring Agriculture with Remote Sensing
MCC	Mesoscale Convective Cluster
MESSR	Multispectral Electronic Self-Scanning Radiometer
MFE	Magnetic Field Explorer
MIEC	Metereological Information Extraction Centre (ESOC)
MIPS	Million Instructions Per Second
MLC	Monitoring Landscape Change
MODIS	Moderate Resolution Infrared Sounder
MODIS-N	Moderate Resolution Imaging Spectrometer – Nadir
MOMS	Modular Opto-electronic Multispectral Scanner
MOS	Marine Observation Satellite
MSFC	Marshall Space Flight Center (Huntsville, AL)
MSG	Meteosat Second Generation
MSR	Microwave Scanning Radiometer
MSS	Multispectral Scanner; Multispectral Sensor
MSU	Microwave Sounding Unit
MTEM	Magnetosphere/Thermosphere Explorer Mission
MTF	Modulation Transfer Function
MWS	Maximum (Sustained) Wind Speed

NASA	National Aeronautics and Space Administration
NDVI	Normalized Difference Vegetation Index
NERC	Natural Environmental Research Council (UK)
NESDIS	NOAA Environmental Satellite Data and Information Service
NEXRAD	Next Radar (operational system, USA)
NIR	Near Infrared
NMC	National Meteorological Center (USA)
NMS	Nimbus-5 Microwave Spectrometer
NOAA	National Oceanic & Atmospheric Administration (USA)
NPOC	National Point of Contact
NPOP	North American Polar Operational Platform
NROSS	Naval Research Oceanographic Satellite System
NRSC	National Remote Sensing Centre (UK)
NSCAT	NASA Scatterometer
NSF	National Science Foundation
OCS	Ocean Colour Scanner
ODNRI	Overseas Development National Resources Institute (UK)
ONR	Office of Naval Research (USA)
OSC	Outer Space Committee (of UN)
PC	Personal Computer
PMI	Passive Microwave Imager
PMR	Passive Microwave Radiometer
POEM	Polar-Orbiting Environmental Monitor (ESA)
PPI	Plan Position Indicator
PRARE	Precise Range and Range-Rate Equipment
PVI	Perpendicular Vegetation Index
QPB	Quantitative Precipitation Branch (of USWB)
R & D	Research and Development
RADAR	Radio Direction and Ranging
RAE	Royal Aircraft Establishment (UK)
RAF	Royal Air Force
RAM	Random Access Memory
RAMP	Radar Mapping of Panama
RAR	Real Aperture Radar
RBV	Return Beam Vidicon
RHI	Range Height Indicator
r.m.s.	Root Mean Square
RPV	Remotely Piloted Vehicle
SAM	Stratospheric Aerosol Measurements
SAR	Synthetic Aperture Radar
SBUV	Solar Back-Scattered Ultraviolet
SCAMS	Scanning Microwave Spectrometer

SCMR	Surface Composition Mapping Radiometer
SCR	Selective Chopper Radiometer
SDD	Stress Degree Day
SDSD	Satellite Data Supply Division
SeaWiFS	Sea-viewing, Wide Field-of view Sensor
SIR	Shuttle Imaging Radar
SIRS	Satellite Infrared Spectrometer
SLAR	Side-Looking Airborne Radar
SLR	Side-Looking Radar
SMMR	Scanning Multichannel Microwave Radiometer
SMS	Stratospheric and Mesospheric Sounder
SMS	Synchronous Meteorological Satellite
SONAR	Sound Navigation and Ranging
SPAM	Spectral Analysis Manager
SPOT	Système Probatoire d'Observation de la Terre
SOM	Space Oblique Mercator
SSM/I	Special Sensor Microwave Imager
SSM/T	Special Sensor Microwave Temperature
SST	Sea Surface Temperature
SSU	Stratospheric Sounding Unit
SVAT	Soil/Vegetation/Atmosphere Transfer
SZ	Subsidence Zone
THIR	Temperature Humidity Infrared Radiometer
TIREC	TIROS Ice Reconnaissance
TIROS	Television and Infrared Observation Satellite
TM	Thematic Mapper
TOMS	Total Ozone Mapper System
TOPEX	Topography Experiment (Ocean)
TOVS	TIROS Operational Vertical Sounder
TREM	Tropical Rainfall Explorer Mission
TRMM	Tropical Rainfall Monitoring Mission
UARS	Upper Atmosphere Research Satellite
UAV	Unmanned Air Vehicle
UKAEA	UK Kingdom Atomic Energy Authority
UNDRO	United Nations Disaster Relief Organization
UNEP	United Nations Environment Programme
UNESCO	United Nations Educational, Scientific and Cultural Organization
USAF	US Air Force
USB	Unified S-band
USDA	US Department of Agriculture
USGS	US Geological Service
USWB	US Weather Bureau
UTM	Universal Transverse Mercator
VDU	Visual Display Units

VHRR	Very High Resolution Radiometer
VIRR	Visible–Infrared Radiometer
VIRSR	Visible and InfraRed Scanning Radiometer
VISSR	Visible and Infrared Spin Scan Radiometer
VTIR	Visible and Thermal Infrared Radiometer
VTOL	Vertical Take-Off and Landing
VTPR	Vertical Temperature Profiling Radiometer
WCRP	World Climate Research Programme
WEFAX	Weather Facsimile
WMO	World Meteorological Organization
WWW	World Weather Watch

PART ONE

REMOTE SENSING PRINCIPLES

1.1 GROWTH IN AWARENESS OF ENVIRONMENTAL PROBLEMS

As we enter the twenty-first century, most people are aware of environmental problems and are concerned about the quality of life they can enjoy in their own locality. These feelings are triggered mainly by events affecting the air they breathe, the water they drink, or the open green spaces they are accustomed to using for leisure. In the main the problems arise from human use of land, air and water, for pressures on land and resources are higher than ever before, and growing fast.

Some people live in constant fear of hazardous environmental conditions, which can lead with frightening speed to disasters that threaten life and property. These environmental threats may involve earthquakes, volcanic activity, flooding, storms, droughts or other factors. Each of these may be of sufficient severity to cause considerable damage to property and loss of human life.

As people become more mobile and see beyond their immediate localities they grow increasingly aware of even larger-scale problems facing particular regions of the global environment. The Earth is the only planet in the universe known to sustain humanoid life. Yet it is human activity that is now conspiring to reduce the life-supporting capacity of planet Earth. In part this is due to the disproportionate consumption of world resources by an affluent minority, and in part to the damage caused by high-tech systems and attendant ways of life. Meanwhile, a majority of the world's population is relatively poor and struggling to achieve better standards of living, including an adequate supply of food. And there is growing evidence that many nations are degrading or even destroying globally significant resources in their own struggle for betterment.

Clearly we as a species have developed the capability of subjugating all other biological species to our own ends. One spectacular result is that the human population of the Earth has grown from about 200 million some 2000 years ago to about 5 billion today. It is still rising today (Fig. 1.1), and population pressures are expected to continue to build in the new century. Therefore, the very important question arises as to when the global population may stabilize. Some authorities have suggested this may not happen until figures of between 10 and 14 billion have been reached.

Through activities resulting from this population explosion we are decimating the biological and genetic diversity of the Earth. An estimated 25 000 plant species and more than 1000 vertebrate species surviving today are now threatened with extinction. Because of these trends, there is justifiable concern that our relationship with the *biosphere* (the thin covering of the planet that contains and sustains life) will continue to deteriorate, and at an accelerating rate, unless a new environmental ethic is adopted.

Moreover, although the human race and other living creatures inhabit the biosphere, the global environmental system contains other important components that interact with it (Fig. 1.2). These include the *atmosphere, geosphere* and *hydrosphere*, each of which makes

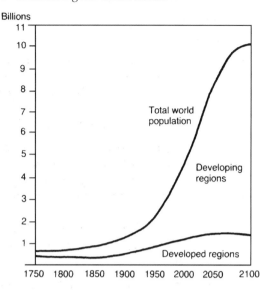

Fig. 1.1 World population growth with future projections.

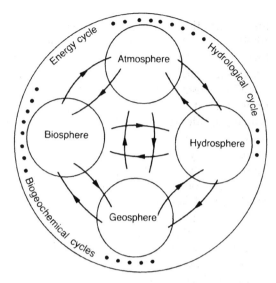

Fig. 1.2 Components of the global environmental system. (Courtesy, NASA.)

important contributions to the conditions of the environment affecting human life. The principal processes influencing the development of the global environment drive the energy, hydrological and biogeochemical cycles which affect us all.

Although it is often hard to dissect out the relationships linking us with our environment that are distinctly natural, technical or cultural, there is no doubt that, through our ability (conscious or unconscious) to interfere with natural phenomena, our species is today a key, possibly *the* key, participant in planetary processes. Many chains of cause and effect are intricately ramified, and inadequately known or understood. The great majority involve so-called *feedback effects*. Thus even our 'improvements' to the environment may prompt another person's environmental problems – or, ultimately even our own.

 Fortunately, alongside a recent growth in environmental concern there has also been a growth in our awareness of the need for *conservation*. Conservation, like 'development', is designed to be beneficial primarily to people. However, whereas development aims to

achieve benefits through exploitation of the environment, conservation aims to achieve benefits by ensuring that such use is 'sustainable'. There is a growing realization that the world economy must be based on *sustainable development*, not the destruction of natural resources. In this respect, living resource conservation has three specific objectives:

1. To maintain essential ecological processes and life-support systems.
2. To preserve genetic diversity.
3. To ensure the sustainable utilization of species and ecosystems.

Many organizations, activities and reports relating to our environment and its everyday use have been evident since the first World Environment Day in 1974. The International Council of Scientific Unions (ICSU) has been a major coordinator of international environmental research programmes. Likewise, initiatives of the United Nations Educational Scientific and Cultural Organization (UNESCO) and the United Nations Environment Programme (UNEP) have played important parts in developing knowledge of environmental problems and how to solve them. Programmes such as the International Geophysical Year (1957–1958) and the International Biological Programme (IBP) running from 1964 to 1974 were early studies of significance. The UNEP has been active in promoting the Global Environment Monitoring System (GEMS), which provides databases for the international community on environmental matters. In more recent years the practical lead in developing an integrated concept of research on the global environment has come from the USA. The National Aeronautics and Space Administration (NASA) and the National Science Foundation (NSF) have made significant contributions to this. One result is the International Geosphere–Biosphere Programme (IGBP), supported by the ICSU. Other large-scale studies of special note are the World Climate Research Programme (WCRP) inaugurated in 1979, and the much younger Global Ocean Observing System (GOOS), confirmed in 1991.

The IGBP is just one of many major projects that are attempting to integrate a wide variety of disciplines and areas of study within global environmental research programmes. In it, particular emphasis is being placed on the need for development of an adequate global data and information system. Without this it will be impossible to make sound policy decisions for the maintenance of a stable and productive Earth environment. In assessing environmental matters, it has become commonplace to follow a well-established sequence of steps. These include:

1. The *recognition* of forms, structures, and/or processes of significance.
2. The *identification* of such phenomena in their real world situation(s).
3. The *recording of their distributions*, often through both space and time.
4. The *assessment of these distributions*, sometimes singly, but more often in some combinations.
5. Attempts to *understand the nature and causes* of any specially significant relationships

between and amongst the key phenomena and processes.

It is only when such background studies have been completed successfully that their results can be used to benefit the human race and its environment. Commonly these results are used in one or more of the following ways:

1. The *preparation and execution of schemes* of environmental or resource management.
2. The *prediction of forthcoming events*, especially those over which we have less direct control.
3. The *planning and development of future projects* to modify and improve the environment.

At the heart of the all-important basic – or background – studies are problems concerned with *observation*, for it is only when we know the true state of our environment that we can predict the future with confidence, including effects of our own impacts on it. Data must be obtained by appropriate means; they must be put into permanent forms; and there are great advantages if they lend themselves readily to computer processing for rapid analysis, interpretation and intercomparison. It is here that environmental remote sensing has its most important part to play, and it is with such activities that this book is mainly concerned. By common consent, remote sensing is beginning to fulfil a long-recognized need for both intensive and extensive monitoring of the Earth, simultaneously in much more detail yet also with much greater uniformity than has ever been possible before.

But first, let us review the development and roles of environmental observation by longer-established means, which will help to set the scene.

1.2 *IN SITU* SENSING OF THE ENVIRONMENT

Humans have made measurements of key aspects of our environment since an early stage in the development of our civilization and culture. Examples of primitive measuring devices include the famous Nilometer by which water levels in the River Nile were noted, and the rudimentary rain gauges used by natural philosophers in the city states of Ancient Greece. The European Renaissance of art, science and literature marked a very significant surge of interest in the need for, and design of, monitoring instruments. Attention began to be paid not only to environmental variables or effects which were readily visible, but also to others less directly evident in the world of nature. In the present context the invention of the thermometer by Galileo at the end of the seventeenth century is particularly noteworthy. A steady deepening of interest in environmental factors and conditions ensued in the eighteenth and nineteenth centuries.

The scientific rebirth of the Renaissance period and its aftermath was followed by a rapid acceleration in our concern for the environment as the twentieth century unfolded. Not only could the environment be monitored by a wide range of sophisticated instruments, but also advances in the related technologies of communications and recording ensured that data could be collected, collated and processed, even from quite remote locations, in 'near real-time' (i.e. very close to the time of observation). At last environmental monitoring could be organized on a wide scale so rapidly that short-term event prediction, and dependent practices of environmental management and control, became possible.

Most recently, advances in other support technologies, especially in the related fields of electronic computing and the microchip, have begun to exercise what will doubtless grow to be a profound influence upon the design of what are often called '*in situ*' sensor networks. Through these, light will be shed on the types of questions we may hope to solve through the analysis and interpretation of the 'conventional' data they proved. Unfortunately, despite great advances in *in situ* technology by the late twentieth century, significant problems in environmental monitoring by such systems remain. Not least significant is the fact that most conventional data are still related to point locations or to transects: for areal assessments, spatial interpolation procedures must be applied. These are more satisfactory when point observation data are more numerous. These remain unsatisfactory for many purposes today because many *in situ* instrument networks are thin, or even virtually non-existent in some regions of the world. In many countries, *in situ* data gathering has actually declined in recent decades, often but not exclusively due to operating economies. Note, too, that many interesting, even important, environmental parameters do not readily lend themselves to *in situ* measurement at all.

Remote sensing, especially from satellites, is rightly seen as at least a partial, and growing answer to all such problems.

1.3 REMOTE SENSING OF THE ENVIRONMENT

Remote sensing can be defined as '*the science of observation from a distance*'. Thus it is contrasted with *in situ* sensing, in which measuring devices are either immersed in, or at least touch, the object(s) of observation and measurement. Remote sensing is applied today in many fields, most notably astronomy, medicine (diagnosis and surgery) and industry (quality control of products) – and in environmental science, our current preoccupation.

Some authors have spoken of remote sensing systems in connection with *in situ* instrument packages which are remote in the sense that they are placed in relatively inaccessible locations, or are connected to central data processing facilities by automatic data acquisition and transmission links. These arrangements do not accord with the definition above, nor with the generally accepted view of remote sensing by those who practise it.

Others have spoken of remote sensing in terms of 'a lack of physical contact' between a sensor and its target. Again, this is incorrect, since where no physical contact exists, no observation is possible: it is true that no 'touch-type' contact exists between a remote sensor and its target, but some physical emanation from, or effect of, the target must be found if aspects of its property and/or behaviour are to be investigated. The most important of the physical links between objects of measurement and environmental remote sensing measuring devices involve *electromagnetic energy, acoustic waves* and *force fields*. These will be discussed in greater detail in Chapter 2.

The home base of our species is located at the interface between the Earth and its atmosphere. For most surface and atmospheric remote sensing, electromagnetic energy is the supreme medium. Therefore, this book is much more concerned with the exploitation of electromagnetic energy than all other energy forms together, and is more concerned with the bottom of the atmosphere and the top of the Earth than with other places or regions in the universe.

In Chapter 2 we will discuss ways in which we all practise naturally some methods of remote sensing: by these we continually observe the world in which we live. But because the science of remote sensing is concerned as much with *recording* as with observation, the remote sensing era may be said to have dawned with the invention of photography in 1826. Devices such as telescopes had been invented much earlier to extend our personal observing capabilities. But it has been only in the last century and a half that means have been devised whereby the environment can be both observed and recorded objectively by artificial devices. In this respect, the potential of photography was quickly appreciated, especially for recording scenes of special significance to the observer. New cameras and types of films were invented, even to extend our view beyond the relatively narrow waveband of visible light. Then began a more systematic search through the radiation spectrum, and into the realms of acoustical, chemical, gravitational and radioactive energy to discover fresh means whereby many other aspects of our natural and cultural environments might be investigated. This search was stimulated greatly by the military demands of two World Wars, especially the second. The search has gathered even more momentum recently, and is not yet complete (see also section 1.5).

It was in 1960 that reference was first made by name to 'remote sensing' either as a distinctive field of study or as a set of approaches to the human environment. Since then it has passed take-off point, deriving great impetus from the opening of the satellite era and the space race between the remote sensing superpowers, the USA and USSR. In particular, NASA has played a gigantic part as a result of American government policy to make remote sensing data (much of it from satellites) readily available to the scientific community throughout the world. The success of NASA has prompted the establishment of national space agencies in other countries, as far afield as Brazil, India and Japan, and international counterparts, most notably in the case of the European Space Agency (ESA), whose Member States are Belgium, Denmark, France, the Federal Republic of Germany, Ireland, Italy, The Netherlands, Spain, Sweden, Switzerland and the United Kingdom. Austria is an Associate Member of ESA, and Canada and Norway have Observer status.

Remote sensing is accredited with a policy-making sub-committee of the United Nations Committee for the Peaceful Uses of Outer Space, a sub-committee more concerned with Earth observation than first appearances suggest: in UN terminology the boundary between 'Inner' and 'Outer' Space is merely 50 km above the surface of the Earth. The research-orientated UN Committee for Space Research (COSPAR) also contributes significantly to environmental remote sensing through its international conferences and working groups.

Opportunities for discussion and exchanges of research results in remote sensing have multiplied dramatically since the early 1970s. National remote sensing societies have been established in most major countries of the world. International links are also growing, for example at a practical level through the European Association of Remote Sensing Laboratories (EARSeL) which is also serving as a catalyst in internationally collaborative research; and at an executive level through CEOS (the Committee on Earth Observation Satellites). This is 'an organization established in 1984 which aims to achieve international coordination in the planning of satellite missions for Earth observation, and to maximize the utilization of data from these missions worldwide'. Spurred on by an associated upsurge of interest in the present and potential applications of remote sensing in everyday operational use, national and international remote sensing centres are also multiplying fast.

Since 1983 a working group of NASA has been examining the mission requirements and major science goals for what is now conceptualized as 'Earth System Science' – the study of the Earth as an integrated, dynamic whole. In this, remote sensing has a key role to play, through a new 'Earth Observing System' (EOS) whose basic structure is set out in Table. 1.1.

In the light of such activity, it is not surprising that remote sensing instruction has already spread widely through the tertiary education

Table 1.1 The recommended programme for a new 'Earth Observing System' as initially conceived by NASA in 1987 (Source: NASA, 1988)

Immediate
Development of new management policies and mechanisms to foster coordination among NASA, NOAA, NSF, others
Strengthening of international agreements and cooperation necessary for world-wide study of the Earth

Near term: 1987–1995	Observing Programme for Earth system science: 1995 and beyond
Continuing and operational Earth observations (NOAA, NASA, others)	New era of integrated, global Earth observations:
Specialized space research missions (ERBE, LAGEOS, UARS, TOPEX/ Poseidon, NSCAT/N-ROSS)	1. Earth Observing System (EOS) on polar-orbiting space platforms (USA, other nations)
Establishment of NASA Earth System Explorer mission series (initiate with GREM)	2. Complementary measurements from the ground and from aircraft, balloons, and ships
Flight of other demonstrated instruments at modest cost	3. Continuation of NASA Earth System Explorer missions (TREM, MFE/ Magnolia, MTEM, GGEM)
Coordinated, interdisciplinary programme of basic Earth system research and *in situ* measurements (NASA, NOAA, NSF, USGS, DoE, ONR, others)	4. Advanced geostationary platforms
	Expansion and utilization of information system for Earth system science
Development of information system for Earth system science	Sustained support for expanded, coordinated, interdisciplinary programme of basic research and *in situ* measurements (NASA, NOAA, NSF, USGS, DoE, ONR, others)
Instrument and technique development	

sector. Here it is often correctly portrayed as the kind of integrative study that is needed in these days to bring together specialists and specialized information from many fields of environmental science, broadly defined. The result is that graduates with interests and expertise in remote sensing of the environment can find employment in such seemingly diverse areas as meteorology, pedology, hydrology, geology and geophysics, agriculture, conservation and protection, pest control, fishery development, land use planning, civil engineering and computing, to mention but a few. Conversely, remote sensing specially welcomes graduates from mathematics, physics, systems and electrical engineering to help it further its own goals.

The needs of the growing community of remote sensing students and scientists are being met by an ever widening range of germane books and journals. The first area-specific journal, *Remote Sensing of Environment*, was published in 1969; the *International Journal of Remote Sensing* first appeared in 1980. Newsletters have proliferated in this field, which is appropriate for one that has been characterized by such rapid growth and change. Some long-established societies have come to terms with the precocious newcomer by taking it under their wings: thus, for example, the *Photogrammetric Engineering* journal has been renamed *Photogrammetric Engineering and Remote Sensing*, and *Transactions on Geoscience Electronics* has become *Geoscience and Remote Sensing*. Internet 'home pages' are the latest area of great growth in information technology (see Appendix).

1.4 ECONOMIC AND PRACTICAL BENEFITS OF REMOTE SENSING, PAST, PRESENT AND FUTURE

It was in meteorology that satellite remote sensing first achieved fully operational status in 1966. In that year the significance of satellite data inputs to meteorological data-pools was acknowledged practically through the inauguration of a system of American satellites designed to yield information to any suitably equipped, relatively modestly priced receiver anywhere in the world – a system within which stand-by satellites were to be available for launching at short notice so that the supply of data to routine users could be reasonably assured, an important prerequisite of any routinely used satellite remote sensing system. That pioneering system has worked well, and almost continuously ever since. Being a relatively self-contained programme, it provided the first convincing figures for benefit/cost assessments of satellite operations. Figures published in 1971 put the annual cost of American meteorological satellite research at around US$200 million, and the annual cost of adverse weather to the American community at some US$10 000 million. Of this huge sum, it was estimated that some 20% could be eliminated given better weather information – equivalent to 10 times the annual expenditure on weather satellites. Already by that time the dramatically increased flow of weather data which satellites had made possible had led to benefits of about US$75 million per annum through improved hurricane forecasting alone – a figure four times the annual outlay on weather satellite operations *per se*.

Later, evaluations of satellite meteorology were made through a study contract of the European Space Agency, relating both to general issues and to the specific issue of the benefit/cost viability of the ESA geostationary satellite system, Meteosat (Chapter 5). At the outset of that study several difficulties were identified, of which some are common to most or all satellite remote sensing studies of Earth phenomena. The most significant of these was that, whilst the costs of satellite remote sensing can be evaluated rather readily, it was more difficult to estimate its returns and benefits. One reason for this was that many 'customers' (e.g. members of the agricultural or construction industries who are advantaged by improved weather forecasts) did not see themselves as users of satellite data and did not pay

challenges are large, and the opportunities for remote sensing careers are growing rapidly.

In every respect environmental remote sensing is here to stay.

1.6 THE GROWTH OF REMOTE SENSING IN GEOGRAPHY AND THE ENVIRONMENTAL SCIENCES

Today the geographical and other environmental sciences make particularly broad and coordinated uses of remote sensing data from many different sources. Figure 1.4 exemplifies the prominent, even pivotal, role that remote sensing is assuming as a data provider and synthesizer in modern environmental research programmes. Therefore it is appropriate that we conclude this chapter with special reference to the growing role of remote sensing in these germane fields. The present high level of interest in remote sensing in geography and other environmental science disciplines can be seen as just the latest stage in a long history of their utilization of data obtained some distance from their physical sources. *Historically*, the use of remotely sensed data in these fields of science can be summarized as follows:

1. *Pre-1925*: a period of slowly increasing study of the use of air-photos in topographic mapping. The use of air survey reconnaissance in the First World War accelerated the process to such an extent that systematic photography from the air became practicable for survey purposes.
2. *1925–1945*: a period of widespread use of aerial photography in photogrammetry and map making. During this span of 20 years, air-photo interpretation became highly developed for military intelligence purposes. Many remote and harsh environments (e.g. Antarctica) were photographed, thus providing information for inaccessible areas.
3. 1945–1955: a period of intense development of (subjective) interpretation and (semi-

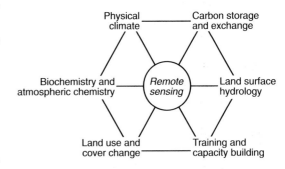

Fig. 1.4 Relations between remote sensing and other research themes within the proposed European contribution to LBA, the Large-scale Biosphere/Atmosphere Programme over Amazonia.

objective) photogrammetric techniques. This was a period when geographers, soil scientists and geologists elaborated on many basic photo-interpretation and photo-analysis schemes.
4. *1955–1960*: a period of extensive application of aerial photography. In addition to topographic mapping, aerial surveys became widely used for regional planning, geological exploration, forest inventory, agricultural studies, terrain analysis, civil engineering, soil mapping and other tasks associated with temporal and spatial changes in the landscape.
5. *1960–1980*: a period of active development of sensing devices, and rapid establishment of satellite platforms, leading to the first fully operational satellite remote sensing systems. The first meteorological satellite, launched in 1960, heralded the opening of this period of intense activity when the potentialities of balloons, rockets and especially satellites as platforms for remote sensing were investigated, using not only conventional photography but also a wide variety of new sensors.
6. *1980–1995*: a period through which many countries other than the USA and USSR (notably France, India, China and Japan) launched Earth observation satellites, and laid plans for greatly increased future

programmes related to remote sensing as a whole. This has also been a period marked by a systematic examination of specific sensor characteristics for particular applications, and intense development of automated image analysis hardware and software packages for computer analysis of remotely sensed data on a 'distributed' (locally effected) as distinct from a more 'centralized' (national or regional) basis.

7. 1995–?: the opening of the period in which commercialization of satellite remote sensing will increasingly help to widen its everyday use – as well as the range of satellite sensors at its disposal.

Geographers in particular have long sought to synthesize information relevant to a particular area or region; often such syntheses have used mapped overlays of different sets of data. Such types of syntheses are now made easier through another relatively recent development, namely that of *Geographic Information Systems* (GIS), which use computer hardware and software to do more efficiently much that was previously done by hand. A Geographic Information System is designed to store large amounts of data and make them available on demand. Although initially computer-based systems for handling geographic information were little more than digital mapping systems, they are becoming increasingly sophisticated today. A classification of GIS functions is shown in Table 1.5. Remote sensing data (as we will see in more detail later) are well suited to such hardware and software technologies. Thus remote sensing and GIS have recently helped to stimulate the development of each other.

Meanwhile, many end-users of these systems have come to expect nearly instantaneous responses to even relatively complex requests. The search is on, therefore, for new methods of simultaneously storing vast quantities of data from both remotely sensed sources (Fig. 1.5) and conventional surveys. New methods of data integration are being

Table 1.5 Classification of GIS functions (NERC, 1989)

Feature	Function
Data input and encoding	Data capture
	Data validation
	Interactive editing
	Quality control
	Data storage and structuring
	Remote access
Data manipulation	Conversion (e.g. vector to raster)
	Geometric correction (e.g. map registration, overlays)
	Generalization and classification
	Enhancement
	Abstraction
Data retrieval	Data selection criteria
	Browse facility
Data analysis	Spatial analysis
	Statistical analysis
	Measurement (e.g. line length, area)
	Error analysis
	Extension into three spatial dimensions and time
Data display	Graphic display
	Map generation
	Text
	Report writing
Database management	Database organization
	Multi-user access
	Maintenance and security

sought using rapid access techniques. Combined *Data Base Management Systems* (DBMS) are seeking to provide ever faster access to spatial data from increasingly large data collection systems. These are vital, for remote sensing data collection systems are yielding exponential increases in new environmental data.

Already commercial GIS and DBMS systems are being marketed for such disciplines as land evaluation, hazard monitoring, watershed management, environmental management, crop yield prediction and urban analysis. It is in such directions that remote

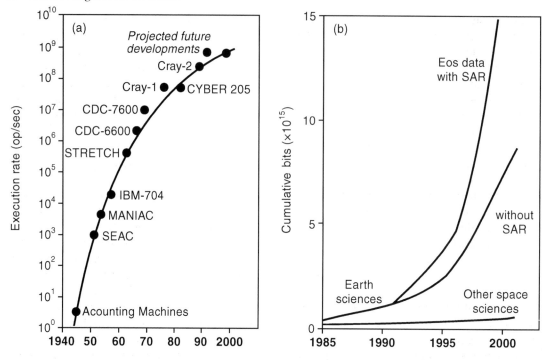

Fig. 1.5 (a) Trends in computational capacity demonstrate the advancing power of supercomputers; the cost per unit computation has fallen steadily, and is expected to continue doing so. (b) In spite of technological advances, the dramatic increase in the volume of data to be processed, validated and made readily available over the next decade (from EOS) will greatly challenge the ground segment. The actual EOS data rate will depend primarily on the duty cycle of the SAR. (After NASA Earth System Sciences Committee, 1988.)

sensing, supported by GIS, needs to develop most if that young science is to fulfil its ultimate promise as an indispensable thought-child of the modern world. For many years remote sensing was dismissed by its critics as 'a set of solutions in search of problems'. As both the detail and the immediacy of data available from remote sensing systems continue to increase, the pertinency of that proposition will continue to decline.

1.7 REMOTE SENSING IN THE NEAR-TERM FUTURE

Hopefully, therefore, the next decade will see remote sensing 'come of age' as a mature discipline with a balanced development of theory, practices and truly operational applications.

The extent to which this will take place in the context of the environmental sciences will depend on a number of factors. The principal one may be the ability of organizations to accept differing data formats as one sensor generation succeeds the other, with different spectral and spatial resolutions, making consequent demands upon database management systems. But other difficulties will also hinder the pursuit of world-wide routine use of remote sensing data, especially difficulties of *technology transfer*. The principal barriers for remote sensing technology transfer to developing countries (amongst those parts of the world which most need, and stand to benefit from, remote sensing) include the following:

1. High costs of data if supplied from elsewhere.

2. High costs of data processing systems for computer interpretation, even though these costs are declining rapidly.
3. Lack of personnel experienced in digital remote sensing and GIS.
4. Difficulties and high costs of maintaining hardware/software systems, once acquired.
5. A shortage of foreign exchange to pay for remote sensing data, data processing equipment and technical support.
6. Lack of knowledge and/or interest amongst decision makers in developing countries as far as remote sensing is concerned.
7. Uncertainties concerning the future of Earth observing satellites provided by the major remote sensing nations.

The first four of these barriers may be said to apply equally to many so-called developed nations also. In particular there is a lack of experienced personnel at higher management levels in the 'user' agencies. As a result the introduction of remotely sensed data into operational administrative systems is taking place more slowly than the facts deserve. Nevertheless, training and processing opportunities are steadily expanding. In the UK, for example, a British National Space Centre (BNSC) was set up in 1985. This provides a wide range of facilities, including some general-purpose high technology data processing and analysis systems. In addition it provides a browse file of available satellite images and a library of books and periodicals dealing with various aspects of remote sensing. Similar national remote sensing centres have been adopted as National Points of Contact (NPOC) in other countries for the dissemination of satellite data from European and US satellites. Furthermore, many countries are seeing the need for regional data processing centres as well as a national centre. These centres are usually located in environmental agencies or departments in universities or polytechnics and act as service points for diferent disciplines or in various regions, offering facilities for training and use of data analysis equipment.

Thus, the use of remote sensing is continuing to grow as an operational tool for environmental mapping, monitoring and management throughout the world, providing both answers to problems, and, where appropriate, profits to company shareholders. It is hoped that, through reading this and other books on the subject, and by visiting websites listed in the Appendix, you will want to play an active part in helping to encourage this, to the benefit of us all, plus the planet on which we live.

2.1 NATURAL REMOTE SENSING

We all use our natural senses to observe and explore the environment in which we live. Certain of these – *smell, taste* and, more often than not, *touch* – permit us to assess environmental qualities directly, through our neurophysical responses to the gases, liquids and solids with which we have immediate contact. The others – *sight* and *hearing* – can make us aware of more distant features through the patterns of energy propagations associated with them. *Feeling*, e.g. manifested through the sensitivity of skin to heat, also enables us to assess some characteristics of distant phenomena. These last three appreciations of the behaviour of energy sources some distance from ourselves are *natural forms of remote sensing*.

Since our visual powers are perhaps the most valuable we possess for gathering information about an object or phenomenon with which we are not in direct contact, we may usefully examine them in more detail as an introduction to the principles involved in remote sensing by *artificial* means. The key components of the remote sensing system which makes sight possible are the eye and the brain. Visible energy, in the form of light emitted by, or reflected from, an illuminated object is detected by sensitive cells in the eye (the sensor). The eye is linked by the optic nerve (cf. a modem) to the brain, which is a high-speed, real-time (i.e. near instantaneous) data processor.

The human eye is sensitive both to the *intensity* of the light energy received and to the *frequency* of the wave-like perturbations which may be taken to characterize such flows of energy. As a consequence we can differentiate both degrees of brightness and a variety of colour tones, respectively. In the brain the quickly-processed data are compressed and experienced as visual images. The brain also serves as a data bank in which earlier images can be stored, albeit within frustratingly narrow time limits, and with considerable loss of accuracy and definition. However, some qualitative 'pattern matching' can be carried out by drawing mental comparisons of, say, the present image and selected images recalled from the past.

Our mental pictures can be modified by simple means – including optical lenses to correct vision defects, selective filters such as polaroid sunglasses to reduce the glare of strong sunlight, and/or (less predictably) by chemicals such as hallucinatory drugs – to influence the way we perceive the area of illumination.

Clearly, a natural remote sensing complex such as the eye–brain system is adapted best to the instantaneous assessment of environmental patterns, the purpose for which it is primarily used. In the context of reflex responses to external stimuli, this would be one of its greatest strengths. On the other hand, where conscious effort and freedom of choice are dominant, this adaptation is one of its primary weaknesses, for in this context it lacks many potentially useful capabilities. These include the following:

1. Both *broader*, and *more selective, abilities to detect variations in environmental conditions:* spectrally our eye–brain system is restricted to the visible region of the electromagnetic spectrum.

future, the simplicity and much lower cost of 'bottom upwards' acoustic echo sounder systems should ensure their continued development and use, with some potential uses in locally based operational meteorology.

2.2.3 ELECTROMAGNETIC ENERGY

Undoubtedly the most important basis for environmental (i.e. Earth surface/lower atmosphere) remote sensing is *electromagnetic radiation*. This is the only form of energy transfer that can take place through free space as well as through a medium. It exhibits enormous variety in its behaviour, and can be exploited by remote sensing in many different ways. So important is radiation energy that some authorities have considered it to be the only basis of significance for environmental remote sensing. Though untrue (sections 2.2.1 and 2.2.2), more than one account of remote sensing contains a categorical statement such as this: 'In remote sensing, information transfer from an object to a sensor is accomplished by electromagnetic radiation.' From section 2.3 onwards, electromagnetism will be our chief focus as a medium for remote sensing, too. This is justified by the strong preponderance of present interest and practice in this field. However, it should be remembered that other modes of information transfer certainly do occur, as summarized above.

2.2.4 ACTIVE REMOTE SENSING

A practical distinction is commonly drawn between 'passive' and 'active' remote sensing. In passive remote sensing (exemplified earlier by the eye–brain system) the sensor detects the energy or force that emanates from the target itself, or some 'third party' source. Most environmental remote sensing systems rely upon these approaches. On the other hand, in active remote sensing the sensor detects signals that are artificially produced. Since such signals are generated under relatively controlled conditions, more can often be learned from the ways in which they are affected by objects in the environment, or by the media through which they are transmitted. RADAR (Radio Direction and Ranging) is a common example of an active system exploiting electromagnetic radiation. SONAR (Sound Navigation and Ranging) is an active system utilizing acoustical energy. Both are so widely used that their acronyms are usually written with lower case letters, a convention we will follow from now on.

2.3 ELECTROMAGNETIC ENERGY

Energy is the ability to do work. It can exist in a variety of forms, including chemical, electrical, heat and mechanical energy. In the course of work being done, energy must be transferred

Table 2.1 Key terms associated with electromagnetic radiation

Processes and phenomena	Entities	Processes	Properties[a]	Behavioural characteristics[b]	Performances[c]
Suffixes	-er or -or	-ion	-ivity	-ance	
Terms	Emitter/Radiator	Emission	Emissivity	Emittance	Exitance
	Absorber	Absorption	Absorptivity	Absorptance	Absorptive power
	Reflector	Reflection	Reflectivity	Reflectance	
	Transmitter	Transmission	Transmissivity	Transmittance	
		Extinction			

[a] Usually related to theoretical (model) maxima on 0–1 scales. Each observed value can be viewed as a natural limitation on the associated performance.
[b] Expressing actual source, medium, or target performance as a ratio of the possible maximum. Sometimes expressed as a percentage (e.g. albedo (the percentage reflectance of natural objects)).
[c] Expressed as radiant fluxes per unit area.

Table 2.2(a) Quantities and units of special significance to remote sensing

Quantity	SI unit	Quantity	SI units
Length (l)	Metre (m) or micrometre (μm)	Area (A)	Square metre (m^2)
Time (t)	Second (s)	Volume (V)	Cubic metre (m^3)
Mass (m)	Kilogram (kg)	Frequency (υ)	Hertz (Hz) or cycles s^{-1}
Temperature (T)	Kelvin (K)	Wavelength (μ)	Metre (m)
Plane angle (x)	Radian (rad)	Incidence angle: angle	Degree (°)
Solid angle (Ω)	Steradian (sr)	between line of sight	
Force ($m\,l\,t^{-2}$)	Newton (N)	and vertical (θ)	
Energy ($m\,l^2\,t^{-2}$)	Joule (J)	Grazing or depression	Degree (°)
		angle: angle between	
		line of sight and	
		horizontal (γ)	
		Density (D)	Kilogram per cubic metre (kg m^{-3})

from one body or one place to another. Such transfers are effected by:

1. *Conduction*. This involves atomic or molecular collisions.
2. *Convection*. This is a corpuscular mode of transfer in which bodies of energetic material are themselves physically moved.
3. *Radiation*. This is the only form in which electromagnetic energy may be transmitted through either a medium or a vacuum.

It is this third type of transfer with which we are primarily concerned in remote sensing studies. Tables 2.1 and 2.2 introduce some key terms, concepts, quantities and units and related terminology.

In keeping with many other areas in the physical and environmental sciences, remote sensing uses *models* (simplified or idealized representations of reality) to describe complex phenomena, situations or interactions. In the case of electromagnetic radiation, two models are necessary to describe and elucidate its most important characteristics:

1. The *wave model*. Any particle with a temperature above absolute zero vibrates. This vibration sets up wave-like disturbances in the electric and magnetic fields that surround the particle, which is acting as a source of radiation. These are perpendicular to each other (Fig. 2.2); furthermore, the direction of motion of these wavy disturbances is always at right angles to them at any time. The wavy disturbances travel away from the source at a common and constant speed, the 'speed of light' (almost exactly 3×10^8 m s^{-1}). The spacing between successive wave crests is the *wavelength* of the radiation; the number of crests passing a selected point in one second is its *frequency*. The two types of wave disturbances vary in phase with each other (i.e. changes in the wavelength of amplitude of the one are matched by changes in the wavelength and amplitude of the other).

2. The *particle model*. This emphasizes aspects of the behaviour of radiation which suggest that it comprises many discrete units, called 'quanta' or 'photons'. These carry from the source some particle-like properties, such as energy and momentum, but differ from all other particles in having zero mass at rest. It has been hypothesized consequently that the photon is a kind of 'basic particle'.

Whilst the true nature of electromagnetic radiation remains obscure, we know that it results whenever an electrical charge is generated. In terms of the wave model, the wavelength (λ) of the resulting ray of energy is determined by the length of time that the charged particle is accelerated; the frequency (f) of the radiation waves depends upon the

2.4 RADIATION AT SOURCE

2.4.1 THE GENERATION OF RADIATION

An electromagnetic source, whether natural or artificial, may emit:

1. a broad continuum of wavelengths of radiation;
2. radiation within a narrow (single spectral) band; or
3. radiation of a single wavelength.

The *intensity* of such radiation *varies with the square of the peak amplitude of the electric field, and is proportional to the number of photons in the field*. Variations in the intensity (and perhaps also the wavelength or frequency) of radiation from a source may occur through time. The electrically charged particles involved in generating the emitted radiation are basic units of matter, namely atoms, electrons and ions. All are in a state of constant motion under normal circumstances, i.e. when the temperatures of the objects they constitute are above absolute zero. Since the molecules of these objects are built of electrically charged particles they possess natural 'resonance', either vibrational or rotational, so that they act as small oscillators which accelerate the electrical charges. The emitted photons of energy have frequencies determined by the resonances of their sources. Note, too, that the distribution of molecules among energy states is a function of temperature, so that at higher temperatures there are proportionally more molecules in the higher energy states. This is to say that not all particles increase their energy equally as the temperature of a radiation source rises, but the overall effect is an increase in molecular activity, accompanied by both an increase in emitted energy and a related shift in the dominant wavelength of radiation towards the higher frequencies.

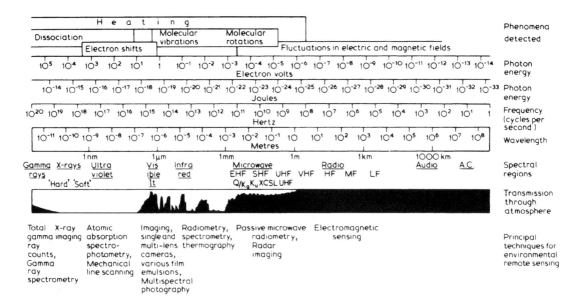

Fig. 2.3 The electromagnetic spectrum. The scales give the energy of the photons corresponding to radiation of different frequencies and wavelengths. The product of any wavelength and frequency is the speed of light. Phenomena detected at different wavelengths are shown, plus the principal techniques for environmental remote sensing. The significance of the results for different environmental applications constitutes a greater part of the latter chapters of this book; students might usefully tabulate them for themselves.

2.4.2 EMISSION OF RADIATION

We have seen that all bodies with temperatures above absolute zero generate and send out, or emit, energy in radiant form. Each radiation source, or 'radiator' (whether natural or artificial), emits a characteristic array of radiation waves which can be assessed in terms of their wavelengths and intensities. For many real-world radiators their array is very complex, comprising a number of different contributions from the various constituents. Hence a characteristic curve, or *spectral emission signature*, may be sought for each type of natural remote sensing target by plotting the intensities of the emitted radiation against appropriate wavelengths in the electromagnetic spectrum. Chapter 3 provides illustrations of a number of such signatures. For the present we must concern ourselves more with the principles that cause such signatures to differ.

A useful concept, widely used by physicists in radiation studies, is that of the *black body*, a model (perfect) absorber and radiator of electromagnetic radiation. A black body is conceived to be an object or substance that absorbs all the radiation incident upon it, and emits the maximum amount of radiation at all temperatures. Although there is no known substance in the natural world with such a performance, the black body concept is invaluable for the formulation of laws by comparison with which the behaviour of actual radiators may be assessed. These laws include the following:

1. *Stefan's* (or *Stefan–Boltzmann's*) *law*. This states that the total emissive power of a black body is proportional to the fourth power of its absolute temperature (*T*). This can be expressed as:

$$M = \sigma T^4 \qquad (2.4)$$

where *M* is radiant exitance, in watts m^{-2}, and σ is the Stefan–Boltzmann constant. This relationship applies to all wavelengths shorter than the microwave: in the microwave region radiant exitance varies as a direct function of *T*. The chief practical

significance of Stefan–Boltzmann's law is that *hot radiators emit more energy per unit area than cooler ones.*

2. *Kirchhoff's law*. Since no real body is a perfect emitter, its exitance is less than that of a black body. Clearly it is often useful to know how the real exitance (*M*) of a radiator compares with that which would be anticipated from a corresponding perfect radiator (*M*$_b$). This may be established by evaluating the ratio M/M_b, which gives the emissivity (ε) of the real body. Thus, for the general case, we may say that:

$$M = \varepsilon M_b \qquad (2.5)$$

The emissivity of a real black body would be 1, whilst the emissivity of a body absorbing none of the radiation upon it (not unexpectedly known as a 'white body') would be 0. Between these two limiting values, the 'greyness' of real radiators can be assessed, frequently to two decimal places. If we plot emission curves for real radiators against curves for corresponding black bodies, we usually find that the *idealized curves are much smoother and simpler than the observed patterns*. In effect the emissivity index is a measure of radiating efficiency across the spectrum as a whole.

3. *Wien's (displacement) law*. This states that the wavelength of peak radiant exitance (λ_{max}) of a black body is inversely proportional to its absolute temperature (*T*):

$$\lambda_{max} = C_3/T \qquad (2.6)$$

C_3 is a constant equal to 2897 μm K. This equation tells us that, *as the temperature of a black body increases, so the dominant wavelength of emitted radiation shifts towards the short wavelength end of the spectrum*. This can be exemplified by reference to the Sun and the Earth as radiating bodies. For the Sun, with a mean surface temperature of about 6000 K, $\lambda_{max} = 0.5$ μm. For the Earth, with a surface temperature of about 300 K on a warm day, $\lambda_{max} = 9.6$ μm. Empirical observations have confirmed that the wave-length of maximum radiation from the Sun indeed falls within the visible waveband of the electromagnetic spectrum whilst that from the Earth falls within the infrared. We experience the former dominantly as light, and the latter as heat.

4. *Planck's law*. This more complex statement describes accurately the spectral relation-ships between the temperature and radia-tive properties of a black body. In one form, Planck's law may be expressed by

$$M_\lambda = C_1 \lambda^{-5} / (e^{C_2} / \lambda^T - 1) \qquad (2.7)$$

where *M* is the energy emitted in unit time for our unit area within a unit range of wavelengths centred on λ, C_1 and C_2 are uni-versal constants, and e is the base of natural logarithms (2.718). A useful feature of Planck's law is that it enables us to *assess the proportions of the total radiant exitance which fall between selected wavelengths*. This can be useful in remote sensor design, and also in the interpretation of remote sensing observations.

In consequence of these four laws we see in Fig. 2.4 that emission curves are characteristi-cally negatively skewed, with their peak inten-sities falling in the lower quartiles of their spectral bands. The wavelengths of maximum emission are progressively shorter for the hotter radiators, and the total emissions from the hotter radiators are greater than those from the cooler radiating bodies.

2.5 RADIATION IN PROPAGATION

2.5.1 SCATTERING

It is unfortunate (for much, but not all, remote sensing) that considerable complications arise in practice where the Earth's atmosphere

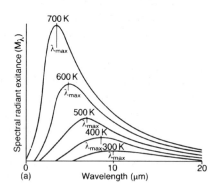

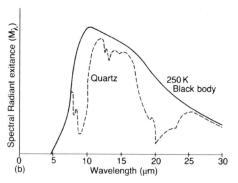

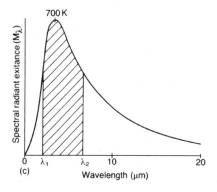

Fig. 2.4 (a) Selected black body radiation curves for various temperatures. In each case the spectral radi-ant exitance (M_λ) is represented by the *y*-axis; total radiant exitance (*M*) is given by the area under each curve. Note that the areas under the curves dimin-ish as temperature decreases. Note too that λ_{max}, the wavelength of peak radiant exitance, shifts towards the longer wavelengths with decreasing black body temperature. (b) The total radiant exitance of a real body (e.g. quartz) is less than that of a black body at the same temperature. The general relationship is expressed by Kirchhoff's law. (c) One form of Planck's law of radiation permits the assessment of that portion of the total radiant exitance falling between selected wavelengths, e.g. λ_1 and λ_2.

intervenes between the sensor and its target. Although the speed of electromagnetic radiation is unaffected by the atmosphere, this medium may affect several of the other characteristics of this form of energy propagation, including:

1. the *direction* of radiation;
2. the *intensity* of radiation;
3. the *wavelength and frequency* of the radiation received by a target, at the base of the atmosphere;
4. the *spectral distribution* of this radiant energy.

Frequently both the direction and intensity of radiation are altered by particles of matter borne in the atmosphere. These redirect, in a rather unpredictable way, the radiation *en route* through a turbid medium. An understanding of radiation scatter is necessary for the selection of sensors or filters where certain effects are required, e.g. where image degradation owing to atmospheric impurities should be avoided, or at least reduced, by sensing at the most appropriate waveband(s). The depletion or *attenuation* which results from 'scattering' (i.e. dispersion of radiation) by particles suspended in the atmosphere is related to the wavelength of radiation, the concentration and diameters of the particles, the optical density of the atmosphere (discussed in section 2.5.3) and its absorptivity. The common types of scatter are:

1. *Rayleigh scatter*. This mostly involves molecules and other tiny particles with diameters much less than the radiation wavelength in question. It is characterized by an inverse fourth power dependence on wavelength. Hence, for example, ultraviolet radiation (about one-quarter the wavelength of red light) is scattered 16 times as much. This helps to explain the dominance of orange and red at sunset when the Sun is low in the sky: the shorter wavebands of visible light are cut out by a combination of atmospheric absorption and powerful scattering.

2. *Mie scatter*. This occurs when the atmosphere contains essentially spherical particles whose diameters approximate to the wavelengths of radiation in question. Water vapour and particles of dust are the main agents that scatter visible light.

3. *Non-selective scatter*. Here particles with diameters several times the radiation wavelengths are involved. Water droplets, for example, with diameters ranging commonly from 5 to 100 μm, scatter all wavelengths of visible light (0.4–0.7 μm) with equal efficiency. As a consequence clouds and fog appear whitish, for a mixture of all colours in approximately equal quantities produces white light.

2.5.2 ABSORPTION

This is the retention of radiant energy by a substance or a body. In the real world it involves the transformation of some of the incident radiation into heat, and the subsequent re-emission of that energy at a longer wavelength.

Considering the ideal case, it will be recalled that a black body is a perfect radiator; it is also a fully efficient absorber of radiant energy which is incident upon it. So, just as the *emissivity* of a real body can be defined as an inherent characteristic of its material, expressing the ease with which it gives up energy by radiation, so the *absorptivity* (α) of a real body is an expression of its ability to absorb radiant energy. Logically, for a black body, $\alpha_b = \varepsilon_b$, and both ε and α equal unity. For 'grey' bodies, ε and α are also equal (with values of less than 1) if they are opaque, in which case both absorption and emission are confined to the (physically simple) surface layer. For all other cases $\alpha_b \neq \varepsilon_b$, and modelling the processes involved can be very complicated and difficult.

In the atmosphere (which can be decidedly murky, but is never opaque) radiation absorption takes place not so much at its surface, but in transit. Three gases, namely water vapour, carbon dioxide and ozone, are particularly (in W m^{-1} K^{-1}) of the rate at which heat can pass through a material. This is generally higher for artificial materials than for natural materials, which are relatively poor heat conductors. Therefore diurnal temperature changes observed in rural areas are more purely functions of the upper layers of water, soil or vegetation rather than their whole depths.

3. *Thermal diffusivity (K).* This is a measure (in m^2 s^{-1}) of the rate of change of the temperature within a body or medium. In general, dry surfaces are observed to diffuse temperature changes more slowly than wet surfaces.

4. *Thermal inertia (P).* This is a measure (in W m^{-2} K^{-1} s$^{1/2}$) of the thermal resistance of a material to changes of temperature. Some materials, e.g. dry sandy soils, have low thermal inertias, and exhibit wide diurnal ranges of temperature. Others, e.g. wet clay soils, have high thermal inertias, and their observed diurnal temperature ranges are relatively low.

The thermal properties of some common constituents of the environment are shown in Table 2.4.

Lastly, reference must be made to rates of heating or cooling. Surface temperature changes prompted by insolation heating during the day and outward radiation by night are important for intelligent uses of brightness temperature patterns and their changes, but can be very difficult to accommodate in brightness temperature data analysis and interpretation because they are influenced by so many factors. These include angles of insolation, slope and aspect, as well as the distributions of screening bodies, such as clouds, trees and buildings. Such factors greatly complicate the task of objective classification of thermal imagery for, although many features of the natural world can be modelled, all models are simplified representations of real-world conditions. For example, in remote sensing studies of vegetation the observed radiation or brightness temperatures are likely to be affected by both leaf and underlying soil temperatures, which depend in turn upon the closeness (or otherwise) of the vegetation cover, the moisture content of the leaves and the soil, the slope of the terrain, the time of day and the season of the year, plus the types and specifications of the sensors and platforms from which the remote sensing observations have been made.

2.8 GENERAL CONCLUSIONS

Although it may have seemed at the beginning of this chapter that Earth scientists wishing to employ remote sensing techniques to improve their perception and appreciation of the human environment need do no more than choose the most appropriate instrument and the best platform for their purposes, it must be apparent by now that many factors may influence that choice. In later chapters we shall review some of the more common sensors, sensor packages and remote sensing platforms available for such uses. However, it is clear

Table 2.4 Thermal properties of some surface types: high (H), high/moderate (HM), moderate/low (ML) and low (L) (After Curran, 1985)

Surface type	Thermal capacity (c)	Thermal conductivity (k)	Thermal diffusivity (K)	Thermal inertia (P)
Vegetation	H/M	L	H	H
Dry soil	M/L	M/L	M/L	H
Wet soil	M/L	M/L	M/L	H
Water	H	L	H	H
Urban areas	M/L	H	L	H/M

already that if we are to understand this rapidly expanding field of scientific activity, rather than just know of it, we must try to relate experimental and operational practice to the principles we have examined thus far. Since this might be difficult without further explanation and illustration, Chapter 3 will be more specifically concerned with observed radiation characteristics of selected real-world phenomena, so as to strengthen the bridge between the rather abstract theory that dominated in this chapter and the more technical array of practical possibilities discussed later in this book.

RADIATION CHARACTERISTICS OF NATURAL PHENOMENA

3.1 RADIATION FROM THE SUN

3.1.1 THE SPECTRUM OF SOLAR RADIATION

Solar radiation reaches the outer limit of the Earth's atmosphere virtually undepleted except for the effect of distance. Since the total radiation from a spherical source, such as the Sun, passes through successively larger spheres as it radiates outward, the amount passing through a unit area is *inversely proportional to the square of the distance between the area and the radiation source*. Although the Earth, 93 million miles from the Sun, intercepts less than one 50-millionth of the Sun's total energy output, this intercepted energy is vital not only to many remote sensing techniques (either directly or indirectly) but also to terrestrial life itself.

Observations of the radiation output from the Sun are made difficult by the attenuation effects of the Earth's atmosphere, at or near whose base most measurements of solar radiation traditionally have been made. Clearly the best vantage point for evaluating the spectrum of solar radiation is outside the atmosphere. Although some data from satellite-borne sensors are now available, probably the most complete and reliable evaluation of it is still that obtained many years ago by staff members of the Smithsonian Institute from the elevated observatory at Mount Wilson. Rocket measurements of ultraviolet radiation have been added to extend the spectrum towards the short wavelength end, where solar radiation is almost completely blocked by absorption in the upper atmosphere. Figure 3.1 summarizes the results, including a curve extrapolated to the outer limits of the atmosphere, and corrected for mean solar distance.

3.1.2 THE ATMOSPHERIC ABSORPTION SPECTRUM

We remarked in Chapter 2 that solar radiation is significantly attenuated by the envelope of gases which surrounds the Earth. Consequently the *curve of solar radiation incident on the Earth's surface falls below that for the top of the atmosphere*, as shown in Fig. 3.1. Some of the attenuated radiation is quickly lost to space by scattering processes, whilst the remainder of it is absorbed within the atmosphere itself. Almost all the solar radiation that penetrates the atmosphere and reaches the surface of the Earth is of wavelengths shorter than 4.0 μm, whereas that emitted by the Earth (Fig. 3.1) is principally in the broad waveband from 4.0 to 40 μm. There is, therefore, practically no overlap between these two types of radiation, which are often referred to as 'shortwave' and 'longwave' radiation respectively. The atmosphere has the ability to absorb not only *direct* solar radiation but also *indirect* (back-scattered) solar radiation, plus longwave (terrestrial) radiation. In considering the absorption of radiation by the atmosphere it is more convenient to review the impact that this has upon solar and terrestrial radiation together, rather than separately.

The patterns of absorptivity of different

In Fig. 3.2 we see, below the curves for individual constituents of the atmosphere, the *total atmospheric absorption spectrum*. At the short end of the atmospheric absorption spectrum the atmosphere absorbs almost all the incident radiation. However, it can be seen that the atmosphere is largely transparent in the visible band from 0.3 to 0.7 μm. There and thereafter in the direction of increasing wavelengths we see a sequence of more or less sharply defined strong absorption bands alternating with relatively transparent regions. These transparent regions are the most useful of the so-called *atmospheric windows* in the absorption spectrum. Others (see Fig. 2.3) are found in the microwave region of the spectrum.

Broadly speaking, practitioners of environmental remote sensing whose major interests are in Earth surface features will select atmospheric window wavebands to explore, and avoid those wavebands in which atmospheric absorption is strongly marked, especially if they plan to use high-level (particularly satellite) observation platforms. However, the wavebands of maximum absorption may be chosen preferentially for some kinds of meteorological and climatological studies. For example, some vertical profiling or sounding

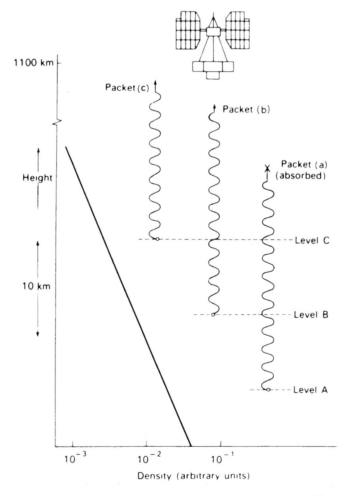

Fig. 3.3 The behaviour of packets of radiation emitted by carbon dioxide from different levels of the atmosphere. (Source: Barnett and Walshaw, 1974.)

of the atmosphere in depth by multiband radiometers focuses upon the 15 μm CO_2-absorption waveband. In studies of this kind the aim is to identify and measure the radiation emitted from a number of discrete levels in the atmosphere, so that linear profiles of the vertical structure of the atmosphere may be obtained (see also Chapter 10).

The basic principle on which satellite depth sounding rests is that the molecules of a gas emit electromagnetic radiation at the frequencies at which they absorb energy. Absorption by a gas such as CO_2 is due to many different vibrational modes of the gas molecules. These modes occur in a series of more or less well-defined bands across the spectrum. So-called *spectroscopic sounding* by satellite-borne multichannel radiometers and spectrometers involves the measurement of radiant emissions from the underlying atmosphere across a series of (often closely spaced) frequencies in each of which the atmosphere absorbs in a known way. Figure 3.3 illustrates this approach through reference to the upward emission of energy of a single frequency. This has been chosen to give a 50% transmittance from level B to the satellite, i.e. radiation packet (b) reaches the satellite half attenuated. Packet (a) is attenuated more on account of its passage through a greater depth of the atmosphere. Packet (c) arrives at the satellite virtually unattenuated, but emission from this level is small: the emission of radiation by CO_2 is proportional to the concentration of the gas, which decreases exponentially with height. Hence the radiation measured by the satellite-borne sensor is a weighted mean of that emitted from various layers, with the greatest contribution from around level B. If upwelling radiation is measured simultaneously at several different frequencies, suitable weighting functions can be calculated to specify the atmospheric pressure levels from which most of the measured radiations originate through the entire depth of the atmosphere (Fig. 3.4).

Radiometric soundings of the atmosphere were pioneered by experimental sensors on research and development (R & D) Nimbus satellites. Operational successors of their sounders, exploiting both atmospheric windows and the intervening 'window frames', are flying now on the US National Oceanic and Atmospheric Administration's polar-orbiting (NOAA) and 'hovering' equatorial (GOES) satellites (Table 3.1). Atmospheric sounding will be discussed further in Chapter 10, in relation to applications of the vertical profile data they provide.

Broad-band 'spectroscopic' sensors pioneered by the Infrared Interferometer Spectrometer (IRIS) of Nimbus have been used to retrieve continuous thermal emission spectra for various environments on the Earth, in addition to vertical profiles of temperature, humidity and ozone concentration, as illustrated by Fig. 3.5 (see also section 4.3.2). IRIS measured planetary radiation in a broad waveband from 6.25 to 22.5 μm using a modified Michelson interferometer. Radiation from the target was split into two approximately equal beams to give an interferogram, which was transmitted to the ground. Here a Fourier transform was performed to produce thermal emission spectra of regions on the Earth (Fig. 3.5(a)). From the water vapour, ozone and carbon dioxide absorption bands in the spectrum, vertical profiles of the concentrations of these gases in the overlying atmosphere were then derived (Fig. 3.5(b)).

Infrared interferometer spectrometers have been used not only to investigate the atmosphere of the Earth, but also the atmospheres of other planets, for example Mars. Still further spectrometers have been tested or proposed, particularly in relation to the greatly enhanced Earth Observation System (EOS) being developed for implementation from the late 1990s onwards, as explained elsewhere in this book. As with many other remote sensing systems designed initially for investigating our terrestrial home, there is a bright future for at least some of them for use in very long-range remote sensing to observe other units in the solar system. Although it is not the purpose of this book to explore such a theme, it is worth noting that considerable progress has already been made with the observation and analysis of the dynamics of atmospheres of other planets in addition to our own.

Figure 3.5 also underlines a general principle of fundamental importance in terrestrial remote sensing, namely that the best spectral regions for observing the Earth on the one hand, and its atmosphere on the other, do not coincide. We must now proceed to explore in greater detail why this is so.

3.1.3 ATMOSPHERIC TRANSMISSION

In view of the almost perfect inverse relationship between the atmospheric absorption spectrum and the spectrum of atmospheric

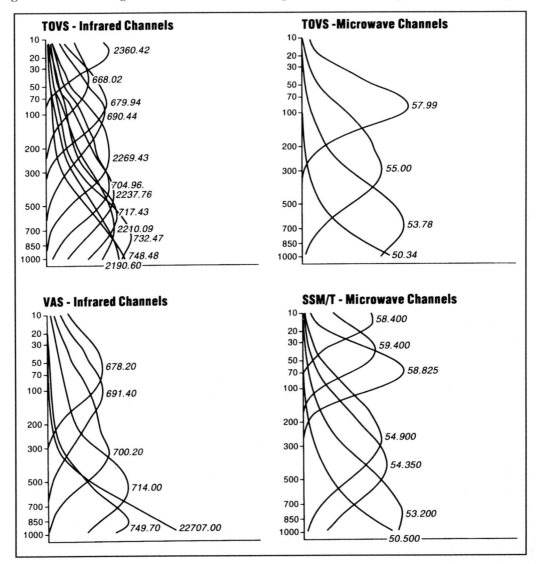

Fig. 3.4 The weighting functions of the atmospheric sounders on NOAA satellites (top left and top right), GOES geostationary satellites (bottom left) and the DMSP satellites (bottom right). Y-axis calibrated in mbar, channel specified in GHz (cf. values in Table 3.1). (Courtesy, NOAA.)

Table 3.1 Channels of the CO_2-dependent High Resolution Infrared Sounder (HRIS-2) and associated water vapour-dependent Microwave Sounding Unit (MSU) carried by operational NOAA polar-orbiting weather satellites in the late 1980s and early 1990s (Courtesy, NOAA)

Channel	Wavelength (µm)	Wave frequency (cm^{-1})	Altitudes of peak emissions (mbar), or general sensitivity
H1	14.96	668.40	30
H2[a]	14.72	697.20	60
H3	14.47	691.10	100
H4[a]	14.21	703.60	280
H5	13.95	716.10	475
H6[b]	13.65	732.40	725
H7[b]	13.36	748.30	Surface
H8[b]	11.14	897.70	Window, sensitive to water vapour
H9	0.73	1027.90	Window, sensitive to O_3
H10	8.22	1217.10	Lower tropospheric water vapour
H11	7.33	1363.70	Middle tropospheric water vapour
H12	6.74	1484.40	Upper tropospheric water vapour
H13[abc]	4.57	2190.40	Surface
H14[a]	4.52	2212.60	650
H15[a]	4.46	2240.10	340
H16	4.39	2276.30	170
H17	4.33	2310.70	15
H18[d]	3.98	2512.00	Window, sensitive to solar radiation
H19[d]	3.74	2671.80	Window, sensitive to solar radiation
H1[e]	0.596[f]	50.30[g]	Window, sensitive to surface emissivity
H2[c]	0.558[f]	53.74[g]	500
H3[a]	0.548[f]	54.96[gl]	300
H4[a]	0.518[f]	57.95[g]	70

[a] Used to compute temperature profiles.
[b] Used to compute cloud fields.
[c] Used in cloud correction.
[d] Used to compute surface temperature.
[e] Used to compute surface emissivity.
[f] in cm.
[g] in GHz.

transmission, the best spectral regions for observing the surface of the Earth are the atmospheric windows in which absorption is low and transmission is high: these are the relative peaks in Fig. 3.5(a), where observed thermal emission spectra most closely approach related black-body spectra. Conversely, the best spectral regions for observing the atmosphere must be the 'window frames' or pillars of high absorption and low transmission: so the profiles in Fig. 3.5(b) were retrieved from the CO_2, O_3 and H_2O absorption bands, respectively, whose positions are indicated in Fig.

3.5(a) (top). Reference to Table 3.1 will confirm that different spectral regions must be exploited for observing the Earth's surface on the one hand and the atmosphere on the other.

Considering these facts, it comes as no surprise to find that a typical atmospheric transmission curve for the atmosphere for radiation wavelengths from 0 to 14 µm (Fig. 3.6) bears a striking upside-down resemblance to the atmosphere's absorption spectrum (Fig. 3.2). However, when we come to consider remote sensing from aircraft (still the mainstay of

(Fig. 3.8). Much current work in remote sensing is now concerned with the compilation of *spectral signatures* derived from such multispectral data, and their interpretation by comparison with 'fingerprint banks' of signatures known to be associated with particular objects or environmental phenomena. Unfortunately there is still much to learn of the spectral responses of many targets in the natural environment, and spectral signature interpretation is often less easy and/or less definitive than some end-users would wish. Many factors affect spectral signatures, including temporal variations such as time of day and season of the year. New multispectral scanners are using many separate channels, and the analysis of the signatures based on large numbers of target responses is almost impossible without the aid of some mechanical or electrical back-up system. For manual analyses, sensors with as few as four carefully selected channels have to suffice to provide maximum contrast for ready qualitative intercomparison.

Since we have already considered introductory examples of remote sensing by the first three of the four means listed above, we may fruitfully turn our attention finally to examples of the fourth. As the analysis of multispectral signatures from the environment is hedged about by many difficulties and uncertainties we should consider first the nature of the responses from some individual constituents of that environment.

3.3 SPECTRAL SIGNATURES OF THE EARTH'S SURFACE

3.3.1 SIGNATURES OF SELECTED FEATURES

It has long been known that rocks, the most fundamental constituents of the solid surface of the Earth, can be distinguished from each other under ideal conditions by their spectral signatures in the thermal emission region of the spectrum (Fig. 3.9). We noted earlier that the emission spectrum from the Sun deviates in some details from the spectrum for a black body with the same mean surface temperature. Similarly many elements of the surface of the Earth have emissivities that vary both in temperature and in frequency, and act more like grey bodies than black bodies. In Fig. 2.4(b), illustrating the distribution of energy emitted by quartz (SiO_2), there were wide differences between the two curves, especially around 9 μm and 20 μm. At these wavelengths incident energy is absorbed sharply by quartz, producing lower rates of emission there. Natural substances often behave more like perfect absorbers and radiators at some frequencies than others, revealing the so-called 'reststrahlen' or residual ray effect, when the actual emission curve is compared with the ideal. Figure 3.9, based on laboratory experiments, confirms that rocks possess the potential to be classified from airborne or satellite sensor data if sufficient spectral detail is generated. Such radiation differences can be observed even from satellite altitudes, for they fall within the broad atmospheric window waveband from about 8 to 13 μm.

Similar differences between spectral signatures of different rocks have been described for the ultraviolet, visible and photographic infrared, based on reflection, rather than emission, spectra. One key source of variability in the observed spectra from particular types of rocks is the water content of the samples. Another is their carbon dioxide content. Reflectance spectra of red sandstone surfaces under different moisture conditions yield different percentage reflections from wet surfaces than those from dry surfaces, especially around 1.4 and 1.9 μm, which are strong water absorption bands. Clearly detailed 'ground truth' data must be obtained if such remote sensing data are to be interpreted correctly (Chapter 6). For many practical purposes this is the type and level of information needed for local uses of remote sensing. Unfortunately, as we shall now proceed to see, such levels of detail are not yet obtainable from satellite systems.

Table 3.2 Aircraft and laboratory studies reveal differences in the percentage reflectances of different types of surfaces and crops. The four bands indicated here are those covered by the Multispectral Sensor (MSS) on the Landsat series of satellites

	Reflectance (%)			
	Band 1 (0.5–0.6 μm)	Band 2 (0.6–0.7 μm)	Band 3 (0.7–0.8 μm)	Band 4 (0.8–1.1 μm)
Rock and soil materials and covers				
Sand	5.19	4.32	3.46	6.71
Loam 1% H_2O	6.70	6.79	6.10	14.01
Loam 20% H_2O	4.21	4.02	3.38	7.57
Ice	18.30	16.10	12.20	11.00
Snow	19.10	15.00	10.90	9.20
Cultivated land	3.27	2.39	1.58	(not given)
Clay	14.34	14.40	11.99	(not given)
Gneiss	7.02	6.54	5.37	10.70
Loose soil	7.40	6.91	5.68	(not given)
Vegetation				
Wheat (low fertilizers)	3.44	2.27	3.56	8.95
Wheat (high fertilizers)	3.69	2.58	3.67	9.29
Water	3.75	2.24	1.20	1.89
Barley (healthy)	3.96	4.07	4.47	9.29
Barley (mildewed)	4.42	4.07	5.16	11.60
Oats	4.02	2.25	3.50	9.64
Oats	3.21	2.20	3.27	9.46
Soybean (high H_2O)	3.29	2.78	4.11	8.67
Soybean (low H_2O)	3.35	2.60	3.92	11.01

Broadening our discussion from rocks and weathered rocks to a wide variety of other environmental constituents, we see from Table 3.2 that many types of soils, crops and other active surfaces, such as ice and snow, are amenable to multispectral identification from airborne or spaceborne sensor systems. Further results from Landsat satellites will be examined in later chapters.

3.3.2 SIGNATURES OF COMPLEX ENVIRONMENTS

A general feature of multispectral images, such as those from the Landsat satellites (e.g. Fig. 3.8), is that they represent the Earth's surface in greater spatial and spectral detail than weather satellites. Although the spectral reflectance properties of some surface constituents can be identified readily by eye, and reported in qualitative terms (e.g. the greater

propensity for water to show differences in reflection from MSS Bands 4 to Bands 5 and 7 than is found over land), more detailed and precise statements are required for many practical purposes. A wide range of automatic and semi-automatic procedures has been developed to meet the resulting need for more objective image analysis, comparison and interpretation. Whilst later chapters will be more concerned with these methods, we remember that many involve some element of spectral signature recognition. However, as Table 3.3 indicates, many influences have a bearing on the spectral signatures which may be derived from multispectral aircraft or satellite data, including the date(s) of imaging, as Fig. 3.10 illustrates.

Often, it is not possible to evaluate all the influences acting on a given set of data. It is rarely (if ever) possible to assess the effects of such influences on each other. Not

Table 3.3 Sources of variation in multispectral signatures of vegetation (Source: Polcyn *et al.*, 1969)

Illumination conditions
Illumination geometry (sun angle, cloud
   distribution)
Spectral distribution of radiation

Site environmental conditions
Meteorologic
Micrometeorologic
Hydrologic
Edaphic
Geomorphologic

Reflective and emissive properties
Spatial properties (geometrical form, density of
   plants, and pattern of distribution)
Spectral properties (e.g. reflectance or colour)
Thermal properties (emittance and temperature)

Plant conditions
Maturity
Variety
Physiological condition
   Turgidity
   Nutrient levels
   Disease
   Heat-exchange processes

Atmospheric conditions
Water vapour, aerosols, etc. (absorption,
   scattering, emission)

Viewing conditions
Observation geometry (scan angle, heading
   relative to Sun)
Time of observation
Altitude

Multichannel sensor parameters
Electronic noise, drift, gain change
Accuracy and precision of measurements on
   calibration references and standards
Differences in spectral responses of systems

surprisingly, therefore, we must note yet again that, contrary to earlier over-optimistic expectations, the *real world is so complex and variable that satellite mapping of environmental patterns is always subject to some doubt and error*: many early promises concerning automatic feature classification were at best premature. Objective feature recognition and mapping is often

more accurate when multispectral data are available, but even multispectral feature recognition and delimitation are often inexact. A further problem which has been recognized relatively recently is that *satellite data often lend themselves best to classifications which do not precisely match those established through large periods of surface observations*. Thus, for example, a distinction is now drawn commonly between (conventional) *land use* data and (satellite-derived) *land cover* analyses. One considerable challenge for the future will continue to be how best to relate *in situ* and remotely sensed data; a second will be to improve methods of combining the two different types of data for best possible use of both.

The complexity of the real world, and multispectral investigations of it, can be well illustrated by reference to a programme carried out over Bear Lake on the border between Utah and Idaho. The left-hand side of Plate 1 (colour section) shows the flight path of the instrument platform, on that occasion a NASA Convair 990 Airborne Observatory. In all, eight sensors were used simultaneously to view the frozen lake and its surrounding snowfields and bare rock. The sensors are summarized in Table 3.4. One viewed in the infrared at 10 µm through the broad atmospheric window. The remainder were all microwave sensors, of which the 1.55 cm sensor was different from the rest in that it was a scanning device, whereas the others all viewed at fixed angles relative to the nadir angle of the aircraft.

Stripchart results from all the sensors are shown in Fig. 3.11. The plot for the 1.55 cm scanner is the average of the five central beam positions across the flight path of the aircraft. Considerable differences are evident from one curve to another. Some of these concern their general forms. For example, the curves of the data from channels 2 and 3 are approximately the inverse of those from channels 5 and 6. Other differences relate to detail embroidered on the basic shapes.

One of the conclusions of this study was

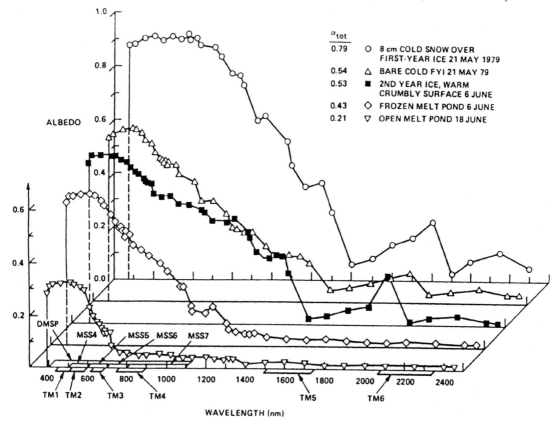

Fig. 3.10 Sea-ice albedos near Barrow, Beaufort Sea, Canada. These values represent samples, as no effort has been made to generalize by season or region. However, a clear progression of albedo spectra is evident over a period of only one month. Some satellite sensor bands are shown (i.e. Landsat, MSS and TM). (Source: Carsey and Zwally, 1986.)

Table 3.4 Characteristics of the radiometers used in a programme of multispectral sensing over Bear Lake, Utah/Idaho, March 1971 (Source: Schmugge *et al.*, 1973)

Frequency (GHz)	Wavelength (cm)	Pointing relative to nadir (deg.)	3dB beam width (deg.)	RMS temperature sensitivity (K)
1.42	21	0	15	5
2.69	11	0	27	0.5
4.99	6.0	0	5	15
10.69	2.8	0	7	1.5
19.35 H	1.55	Scanner	2.8	1.5
37 V	0.81	45	5	3.5
37 H	0.81	45	5	3.5
Infrared	1.0×10^{-3}	14	<1	<1

that at the longer wavelengths (especially 21 cm) the observed variability in brightness temperatures, especially over the lake, was related to thickness variations in the ice. Results indicated that the transparency of ice and snow is a function of wavelength, and that ice and snow depth may be assessed by these longer microwave emissions.

The right-hand side of Plate 1 (colour section) is the 1.55 cm microwave image of the

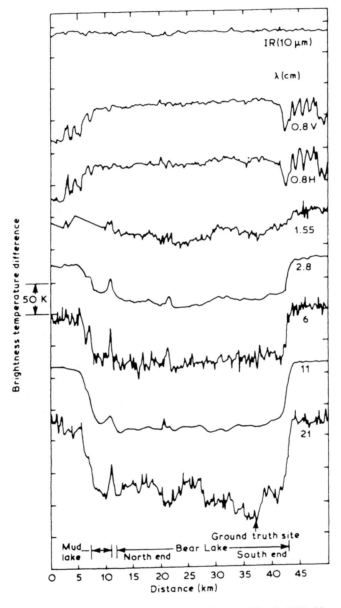

Fig. 3.11 Multispectral data obtained over Bear Lake, Utah/Idaho, 3 March 1971. H and V refer to the horizontal and vertical channels of the 0.8 cm radiometer, which viewed the surface at an angle of 45°. The remaining radiometers were nadir viewing. Each graph is related to the average temperature value at the frequency in question. (Source: Schmugge *et al.*, 1973.)

pass over Bear Lake at an altitude of 3400 m. The area it covers corresponds with part of the flight path. The outline of the frozen, variably snow-covered, lake can be clearly seen. The low brightness temperatures of the frozen surface contrast sharply with the higher temperatures along the steep slopes of the eastern edge of the lake. Although a considerable amount of cloud was present over the target area when the flight was made, useful results were still obtained since microwaves, as we noted earlier, penetrate most clouds.

More recently, multispectral scanning and analytical science have developed very rapidly, involving complex imaging spectrometers (or 'spectroradiometers') and advanced processing techniques: investigating and evaluating the radiation characteristics of specific phenomena and environments simultaneously through a large number of wavebands is an increasingly attractive and effective form of remote sensing. For the most part, operational environmental mapping and monitoring to date have been concerned with the relatively simple – but often only moderately successful – task of analysing integrated radiation or radiation patterns obtained through relatively small numbers of wavebands at one time. Thus there is a new challenge for the future: we know much about the basic physics of radiation reflection and emission, but we still need to learn much more about the often highly complex reflection and emission patterns that are associated with real phenomena and situations.

A good example of a situation that should be clarified by spectroradiometry from space is afforded by phytoplankton, sudden upsurges (or 'blooms') of which occur from time to time, especially in lakes, coastal waters and enclosed seas. Such blooms usually occupy quite small areas (from a few tens of square kilometres to hundreds), last for 10–100 days, and are often nearly mono-specific. Since they are known to encourage significant fluxes of carbon to deeper waters as they decay, their creation and dissipation is now being

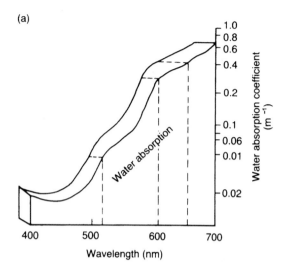

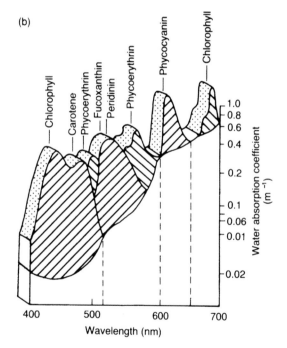

Fig. 3.12 Absorption of light (a) by water and (b) by different algal pigments in the windows of 'clarity' of water; the spectra for the pigments approximate those measured *in vivo*; fucoxanthin and peridin are superimposed. (After Yentsch and Yentsch, 1984.)

urgently sought in relation to primary oceanic productivity, the global carbon cycle, and the possibility of continued global atmospheric warming. Figure 3.12(a) illustrates the common profile of sensible light absorption by water in the absence of phytoplankton; Fig. 3.12(b) shows the increased absorptions associated with common phytoplankton pigments. For example, algae containing chlorophyll (chlorophytes) usually transmit light in the green region, those containing phycocyanin (cyanophyceans) transmit mostly in the blue region, while algae containing chlorophyll a, chlorophyll c (diatoms, dinoflagellates, chrysophytes) and phycoerythrin (cyanophyceans) transmit mostly in the yellow. Sensors such as the Landsat MSS and TM and the Nimbus-7 CZCS provide only enough information to yield spectral signatures with the level of detail in Fig. 3.12(a); to generate the much more complex signatures exemplified by Fig. 3.12(b) spectrometers providing for the simultaneous collection of image in 100 or more continuous spectral bands are required, yielding continuous reflectance spectra for each pixel in the scene.

One such system being developed for the EOS programme is the High Resolution Imaging Spectrometer (HIRIS). Along with planned developments of the NOAA-AVHRR sensor (section 5.9) and MODIS-N (Table 14.2), this will provide data with the spectral resolution necessary to assess plankton populations, as well as other recognition/classification problems of a physically similar nature, e.g. detailed rock and mineral identification, plant classification, and analyses of organic particles suspended in water. The development, testing and use of such systems will become a major thrust of satellite remote sensing of natural phenomena in the late 1990s and beyond. However, interesting challenges may well remain. For example, it is generally recognized that improved corrections are required to

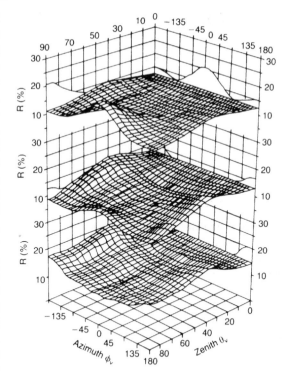

Fig. 3.13 Angular reflectance distributions above savannah at the surface (bottom data) for a clear atmosphere (aerosol free), at the top of a mildly aerosol-loaded atmosphere with surface visual range $V_0 = 50$ km (middle data), and at the top of a heavier aerosol-loaded atmosphere with $V_0 = 10$ km (top data). Sun direction is $\theta = 30°$, $\phi = \pm 180°$, $\lambda = 0.85$ μm. (Courtesy, NASA.)

adjust for atmospheric effects on upwelling radiances measured by satellites, for whatever purpose. Figure 3.13 emphasizes that, in order to observe the very complex effects of the atmosphere on such radiances, account must be taken of the sensor viewing angle (for different responses will be observed from different viewing angles) – and for greatest accuracy simultaneous observations from several viewing angles are required. As yet these are still beyond the capabilities of the sensor systems now being built.

SENSORS FOR ENVIRONMENTAL MONITORING

4.1 INTRODUCTION

In this chapter the sensors available for remote sensing studies will be examined in more detail in order to explain the advantages and disadvantages attached to each sensing system. As Figure 4.1 indicates, the most convenient subdivision for present purposes is between *imagers* (basically providing information in the horizontal plane) and *sounders* (which vertically profile their target media, most commonly air or water).

Two broad categories of passive imaging sensors can be identified, namely *photographic* and *non-photographic*. The former generates a record of radiation from the target to provide

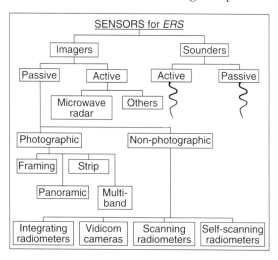

Fig. 4.1 A classification of the more widely used sensors for environmental remote sensing. Several types of sounders exist, but more stress is appropriate here on types of imagers.

'hard-copy' (paper or film) images; the latter provides some kind of 'soft' (electronic) imagery which may or may not be processed to provide hard copy. Photographic systems operate in the visible and near-infrared parts of the spectrum (0.36–0.9 µm) only, whereas non-photographic sensors can range from X-ray to radio wavelengths (Fig. 4.1).

4.2 PASSIVE SENSORS IN THE VISIBLE WAVELENGTHS

4.2.1 PHOTOGRAPHIC CAMERAS

The photographic camera is a well-known remote sensing system in which the focused image is usually recorded by a photographic emulsion on a flexible film base. Photography in various guises still forms the basis of much environmental remote sensing today, and must be explained in commensurate detail.

There are three basic types of camera: *framing*, *panoramic* and *strip* cameras. The framing camera usually provides a square image with an angle of view up to 70°. The panoramic camera gives a very wide field of view which can extend to the horizons on either side, with consequent distortion of the image. There are various types of panoramic camera systems and the commonest type uses a reciprocating, or continuously rotating, narrow-field lens. The strip camera consists of a stationary lens and slit, together with a moving film of width equal to, or greater than, the slit length. The film is moved across the slit at a speed which

5. control of roll, pitch, yaw, vibration;
6. optical quality of any filter or window placed in front of the lens.

The definition of photographic image quality has been quoted conventionally in terms of 'resolving power'. This is still usually measured by imaging a standard target pattern (Fig. 4.3) and determining the spatial frequency (in lines mm^{-1}) at which the image is no longer distinguishable. Resolving power can be determined also by examining the contrast and distortion of the image of a sinusoidal grating object. This measurement, as a function of the spatial frequency of the grating, gives what is termed the *modulation transfer function* (MTF), which is commonly used to express the resolution characteristics of different films. Examples of the sensitivity and resolving power of widely used Kodak aerial films are given in Table 4.1.

Table 4.1 Sensitivity and resolution for selected Kodak aerial films

Type	Sensitivity (aerial film speed)[a]	Resolving power (lines mm^{-1}) at test object contrast	
		1000:1	1.6:1
High-definition Aerial 3414	8	630	250
Panatomic X 3400	64	160	63
Plus X Aerographic 2402	200	100	50
Double X Aerographic 2405	320	80	40
Tri X Aerographic 2403	640	80	20
Infrared Aerographic 2424	200	80	32
Aerochrome Infrared 2443	40	63	32
Ektachrome Aerographic 2448	32	80	40

[a] Aerial film speed for monochrome negative material is defined as $3/2E$, where E is the exposure (in metre-candela-seconds) at the point on the characteristic curve where the density is 0.3 above base plus fog density, under strictly defined conditions given in ANSI Standard PH2.34–1969. A doubling of aerial film speed number denotes a doubling of sensitivity.

A summary of the characteristics of the different types of film available is given in Table 4.2, from which it will be apparent that the versatility of the camera sensing system is increased by the availability of different films with different spectral ranges, sensitivities and resolving powers.

Where images are required at different wavelengths, multispectral photographs may be taken either by mounting several identical cameras with different film emulsions or using a special multispectral camera (Fig. 4.4). In the example shown, several identical lenses record their separate images on a common film. The four-lens system is the most common arrangement; the Itek camera has as many as nine lenses. An example of multispectral photography by the I^{2}S camera system is given in Fig. 4.5. This photograph was obtained with Infrared Aerograph 2424 film using four filters, which provided images at 0.4–0.5 μm (1), 0.5–0.6 μm (2), 0.6–0.7 μm (3) and 0.7–0.9 μm (4). These separate images can be recombined and analysed as discussed in Chapter 7.

There is constant research aimed at improving the sensitivity and resolution of films, whilst reducing their thickness (weight). Sensitivity has been approximately doubled each decade since 1850.

The principal advantages of the photographic camera are its large information storage capacity, high ground resolution, relatively high sensitivity and high reliability.

4.2.2 SPACE-BORNE PHOTOGRAPHIC CAMERAS

The main shortcomings of photographic surveys from spacecraft are that exposures can be made only in daylight, cloud obscures ground detail, and the photographic film cannot be reused. It has been estimated that, if scanning systems had not been used on Landsat, the film weight required to cover one year of filming operations to obtain 60 m ground resolution would be 300 kg. On-board processing would have nearly doubled this weight, and

Table 4.2 Summary of film characteristics (Source: Curtis, 1973)

Film	Disadvantages	Advantages
Panchromatic	Limited tonal range	Sharp definition Good contrast Good exposure latitude Inexpensive
Infrared	Limited tonal range Very high contrast Slight resolution fall-off (without correction for focal distance) Difficult to determine correct exposure Loss of shadow detail	Vegetation evident Tracing of water courses facilitated Inexpensive
Colour	Expensive Special processing facilities necessary Diffusion of image under high magnification and slightly less definition than panchromatic	Excellent all-round interpretation properties owing to good contrast and great tonal range Good exposure latitude when used as a negative film Good-quality black and white prints can be produced from negatives
False colour	Expensive Special processing facilities necessary Diffusion of image under high magnification Critical exposure Low ASA rating Less tonal range than colour Duplicates expensive and difficult from positive film	Sharp resolution Superior rendering of vegetation and moisture

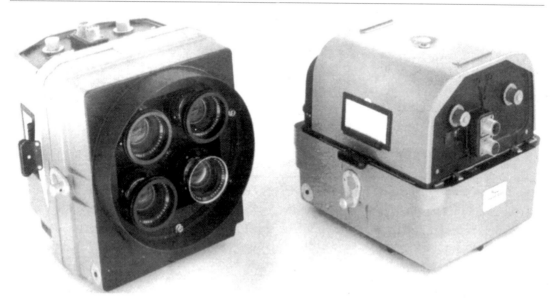

Fig. 4.4 International Imaging Systems (I²S) multispectral aerial camera. Four Schneider Xenotar 150 mm or 100 mm lens. Film capacity 242 mm in width by 76.2 m long. Front view on left, top and side view on right. (Courtesy, John Hadland (PI) Ltd, Bovington, Herts.)

Fig. 4.5 Multispectral photograph of part of Thetford, Norfolk. Apart from the differentiation of deciduous and coniferous trees, Band 4 (infrared) also provides considerable additional detail concerning roof structures. Road details are better portrayed in Band 3 (red) at top left. A combination of Bands 3 and 4 offers advantages in studies of urban landscapes. (Courtesy, NERC.)

improving the resolution to 20 m (the approximate resolution of Landsat TM data) would have increased the film weight to 3000 kg. Such weight factors must be considered alongside the difficulty in transferring the information on the film to Earth. One such method of transfer involves processing the film on the satellite and electronically scanning it for picture transmission to ground stations by broadband telemetry. The other method is by ejecting the film from the satellite for recovery in the air or on the ground. Capsules of 39 kg (air retrieval) and 136 kg (ground retrieval) have been used in this way with military reconnaissance satellites.

The Metric Camera, developed in Europe,

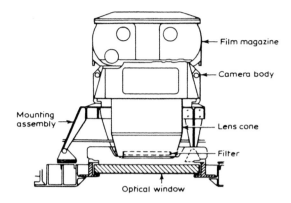

Fig. 4.6 Metric Camera for the first Spacelab pay-load. (Source: European Space Agency.)

and the American Large Format Camera provide examples of recent sensors in the civilian domain which afford high-resolution space photography. The Metric Camera configuration is shown in Fig. 4.6 and was first tested in the 'Metric Camera Experiment' (Chapter 5). Excellent images of the Middle East, North Africa, Pakistan, Afghanistan and West China were at once obtained by this camera.

The Large Format Camera (LFC) was designed in the USA and made its maiden voyage in a Space Shuttle Challenger mission in 1984 (Chapter 5, Table 4.3 and Fig. 4.7). Although designed primarily for mapping, individual photographs have proved useful for many types of qualitative studies for Earth resources studies and environmental

Table 4.3 Operational characteristics for the Large Format Camera (Source: European Space Agency)

Parameter		Specification
Lens		
Focal length		30.5 cm
Aperture		f/6.0
Spectral range		400–900 nm
Image format		22.8 × 45.7 cm
Field of view		
Along track:	degrees	73.7
	altitude ratio	1.5 × H
Across track:	degrees	41.1
	altitude ratio	0.75 × H
Forward overlap modes		80, 70 or 60%

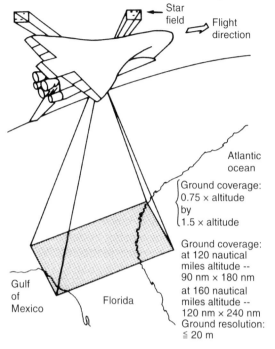

Fig. 4.7 Large Format Camera (LFC) payload ground coverage. (Source: NASA.)

monitoring. For example, studies of the uses of the LFC photographs for coastal mapping showed that high-resolution space photography has a role to play in change detection of coastal features. As the effects of global warming become apparent through raised sea-levels, such observations may be important in planning sea defences. Recent developments in high-resolution digital imagers are now leading to a lessening of interest in the use of more conventional cameras on space platforms, and this trend is likely to accelerate.

The highest resolution photography commercially available in the early 1990s was provided for sale by Soyuzcarta, the Moscow-based agency for Soviet low Earth-orbit imagery. Until recently it has been obtained by the Russian KFA-1000 camera, now superseded by the MK-4: this gives lower resolution (about 63% of that of the KFA-1000) but with greater ground coverage per frame. The ground resolution of the KFA-1000 images is 5–6 m, using later chapters of this book. However, it must be noted that infrared radiometers are also used extensively for local studies from aircraft and helicopter platforms, whether alone or in support of satellite investigations.

The image-forming sensors include infrared linescan and thermal imager systems (section 4.3.2). All are designed to scan a series of lines or narrow strips across the scene so that an image can be built up progressively across the target as in Fig. 4.8(a), where a rotating mirror collects the target radiation, or as in Fig. 4.8(b), where the collector oscillates back and forth. A third linescan type is the conical scanner (Fig. 4.8(c)). This is advantageous because all the data therefrom are obtained the same distance from the satellite, and from the same viewing angle. This ensures that all pixels have the same size on the ground, and eliminates *limb-darkening* effects (the reductions in radiation observed from a target owing to atmospheric effects, which increase with increasing obliqueness of view from the satellite). It will be apparent in each case that scan rate must be related to the altitude and velocity of the sensor when assessing the performance of the system.

Infrared linescanning systems are used extensively from aircraft and helicopter platforms. For example, the Infrared Linescan Type 212 manufactured by Hawker Siddeley Dynamics has a temperature resolution of 0.25°C at 0°C background temperature; thus surface temperatures can be detected to an accuracy of less than 1°C. When detector signals are processed from an image the resultant thermal maps can provide great detail concerning temperature variations. Two examples are shown in Figs 4.9 and 4.10, where the heat from aircraft engines and the warmer areas in the shelter of hedgerows can be easily recognized.

Typically, such airborne infrared linescanning systems generate data which

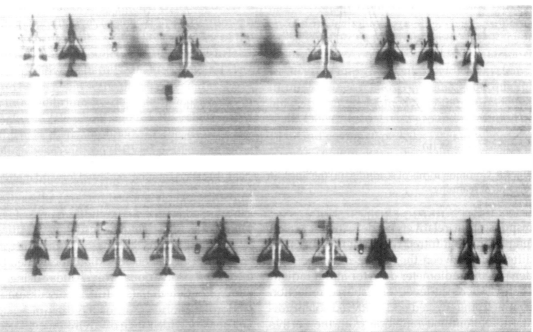

Fig. 4.9 Infrared linescan image of aircraft on an airfield. The heat from engines is shown in light tones. Note also the thermal shadows of recently departed aircraft at the top of the picture. (Courtesy, Hawker Siddeley Dynamics Ltd, Hatfield, Herts.)

Fig. 4.10 Infrared linescan imagery of land near Mark Yeo, Somerset. Note shelter effect of hedges showing in light tones of areas with higher temperature. Grazing animals can also be seen by reason of their high body temperature. (Courtesy, Royal Signals and Radar Establishment, Malvern. Crown copyright.)

indicate relative thermal energy differences between objects scanned. However, the *absolute* difference in temperature can only be estimated in such cases. To achieve absolute temperature relationships from scanner data, several additional constraints must be met. First, the video electronics must have DC (direct current) response to follow faithfully each thermal level variation from the detector. Second, radiation received by the detector must come predominantly from the area of interest being sensed. Last, the complete electronic chain from detector through to final recording must be 'drift stabilized' so that the system response is constant.

The Daedalus DS-1230 is a quantitative scanner fulfilling the absolute temperature requirements set out above. This scanner provides high spatial resolution using a detector size of 1.7 milliradians combined with temperature resolution of 0.2 K.

The linear array or 'pushbroom' system now deployed on SPOT satellites is described in section 5.8. Suffice to say here that this has provided some of the highest resolution imagery from civilian satellite-borne radiometers or spectrometers, down to about 10 m on the ground (Fig. 4.11), and this type of system will be used increasingly in the future.

Meanwhile, *thermal imagers* produce images very similar to those obtained by linescanners, but from sensors in this case scanning in both dimensions. A high quality display can be obtained by using an array of detector elements matched to a suitable scanning system, as shown in Fig. 4.12. This is the so-called *banded scan system*, in which the master scan is generated by a rotating drum with progressively angled facets. This type of sensor is essential for those geostationary satellites where there is no vehicle motion relative to the Earth (Chapter 5). In any case, as scanning height increases, the thermal imager may become preferable to the simple linescan in that it can be matched more efficiently to the scan angle to be covered, and its scanning rate can be controlled to provide the desired compromise between sensitivity, resolution and scanning frame time. This type of sensor also lends itself readily to the use of a range of detector elements, thereby increasing sensitivity for particular conditions of optics and scan rate. The HCMM and CZCS radiometers (sections 5.11.1 and 5.11.3) are examples of earlier space systems of this type.

4.3.2 ABSORPTION SPECTROMETERS

Spectroscopic techniques can be applied to various types of passive sensors leading to instruments commonly referred to as spectrometers. Essentially, spectrometers are instruments used to determine the *wavelength distribution* of radiation (Fig. 3.5). This can be done by dispersing the received radiation spatially according to wavelength. The *prism spectrometer* relies on the dependence of the index of refraction of various prism materials on the wavelength of radiation entering the prism. The *grating spectrometer* achieves dispersion by diffraction and interference. One of the most popular grating spectrometer systems used today was devised by Ebert some 60 years ago. Scanning of the spectrum is achieved by applying a sawtooth type of oscillatory motion to the grating.

A *rotating, circularly variable filter* may replace the prism or grating and thereby gain an increase in the rate at which incoming

Fig. 4.11 SPOT HRV (panchromatic) scene of Athens, Greece, in which not only can the airport be clearly seen, but also individual roads, fields and, in some cases, large buildings. Ships in the harbour area are also clearly visible. (Copyright, CNES.)

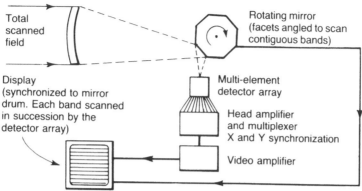

Fig. 4.12 Thermal imager system – block schematic. (Source: Laird, 1977.)

radiation can be scanned. When a circular filter of this kind is used with a rotating beam modulator (chopper) and an internal black-body reference source, calibration of the spectrometer measurements can be achieved (Fig. 4.13). Commercial circularly variable spectrometers are available for airborne use with the capability of interchanging detectors. These instruments allow coverage of waveband frequencies in the range from 0.35 to 23 µm and can accommodate both cooled and uncooled detectors. In the context of airborne remote sensing, Airborne Imaging Spectrometer (AIS) data are now being used in mineral exploration. The AIS has enabled geologists to collect data in a continuous

spectrum and carry out spectral analysis to determine individual absorption band characteristics.

From satellites, infrared spectrometers have been used to obtain atmospheric profiles – generally values of temperature and water vapour as a function of height. The first such infrared sensor that demonstrated the feasibility of remote profiling of the atmosphere was the Satellite Infrared Spectrometer (SIRS), which used an Ebert type grating spectrometer. Today's much more advanced vertical sounders cover many infrared bands (Table 3.1), with high temperature accuracies (better than 1 K), useful relative humidity accuracies (better than 10%), and very good total ozone content sensitivities (about 0.01 cm).

A growing area of absorption spectrometry involves the recognition and measurement of different gases, depending on the unique spectral signature of each gas (Fig. 4.14). At present, spaceborne applications of remote gas sensing have been mainly limited to upper atmosphere studies. However, the principles and instruments described can be applied in many contexts, e.g. the detection of pollutants and emitted gases and vapours from Earth surface sources. The development of space instruments specifically for the detection of Earth pollutants or Earth resources is now being actively pursued. New instruments are being built, and further technological

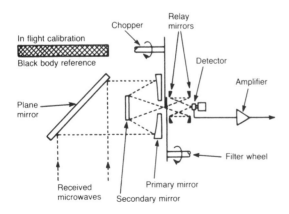

Fig. 4.13 Elements of a filter-wheel radiometer, i.e. spectrometer.

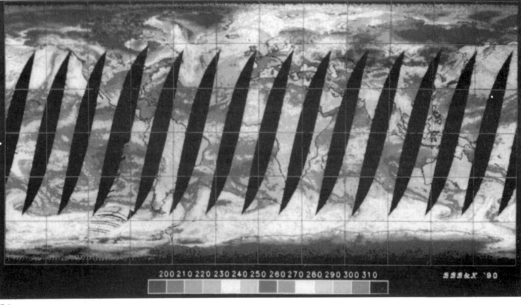

Fig. 4.15 Global mosaics of passive microwave imagery from the 85 GHz (vertical polarization) channel on SSM/I instrument for 20 February 1988. This instrument was first flown in 1987 on the DMSP F8 satellite. DMSP is a near-polar orbiting satellite family permitting (a) night-time imaging along its *ascending* tracks, and (b) day-time imaging along its *descending* tracks. Note the different swath patterns in these two cases, and the higher emissions from land areas especially in the southern (summer) hemisphere by day, and from sea areas in the tropics where atmospheric moisture contents are highest.

Table 4.6 Original expectation for parameters measurable with passive microwave sensors and the frequencies at which the measurements should be made (Courtesy, NASA)

Parameter	Frequency of observation (GHz)									
	1.4	6	10	18	21	37	50–60	90	160	183
Soil moisture	●	○								
Snow		○	○	●		●		◐		
Precipitation										
Ocean			◐	●	○	◐				
Land				◐		●		●		◐
Sea-surface temperature	●	◐	◐	◐	○					
Sea ice										
Extent				●		●		○		
Type	○	◐	●			●		◐		
Wind speed (sea surface)		●	◐	○	○					
Water vapour										
Total (over ocean)		●		●	◐					
Profile				◐	○		◐	○	◐	●
Cloud water (over ocean)				◐	●			◐		
Temperature profile					○	○	●	◐		

Key: ● necessary, ◐ important, ○ helpful.

properties such as temperature or water vapour. For example, the Nimbus-5 Microwave Spectrometer (NEMS), Scanning Microwave Spectrometer (SCAMS) and Microwave Sounding Unit (MSU) were microwave sounders designed to obtain profiles of atmospheric temperatures, although the Nimbus-5 NEMS and SCAMS also obtained ancillary information on features such as rain, snow and ice coverage.

Among the earliest of the imaging passive microwave radiometers was the Electronically Scanning Microwave Radiometer (ESMR), a single-frequency sensor flown in 1972. The successor to the single-frequency sensors was the Scanning Multichannel Microwave Radiometer (SMMR). This was a multifrequency imaging device, with five frequencies, each dual polarized, scanning at a fixed angle of about 50° at the Earth. Today's successor is the conically scanning Special Sensor Microwave Imager (SSM/I) on USA military satellites (see Fig. 4.15 and Chapter 10).

Tables 4.5 and 4.6 confirm that for many applications it is essential to make measurements in multiple frequencies. Multiple

frequency microwave imaging of the Earth's surface (Fig. 4.16) has proved successful because, by analogy with multispectral photography, the increased information provided by simultaneous measurements at different frequencies makes it possible to detect additional phenomena and to isolate relevant data from other surface or atmospheric radiation effects by such means.

Although naturally emitted microwave radiation intensities are much lower than those in the infrared (Fig. 3.1), resulting in poorer brightness temperature resolution, the longer wavelengths have the important advantage that they *allow sensing through cloud cover*.

The microwave radiometer is made up of a directional aerial, a receiver (for selection and amplification) and a detector. The ground resolution achieved by microwave radiometers depends on the aerial type and size, and the orbiting altitude. Since it is important that aerial size should be limited in order to avoid undue weight or destabilization of the satellite, where conventional aerials are planned the lowest possible altitude orbit should be used for maximum resolution.

The principal characteristics of passive microwave radiometer (PMR) systems are, therefore:

1. a relatively low spatial resolution capability, usually of the order of 10 km or more;
2. a wide range in the observed surface emissivity;
3. nearly all-weather capability.

As in other situations, it is convenient when considering the measurements made by microwave radiometers to use the concept of brightness temperature, T_B (p. 33).

Both natural and man-made surfaces present a large range of microwave brightness temperatures. Whereas for a black body the brightness temperature is equal to the physical temperature and ε is unity for all incidence angles, natural objects (grey bodies) show variations in emissivity depending on incidence angle, wavelength and the degree of surface roughness. Where surfaces are very rough the brightness temperature is independent of angle of incidence, but if roughness is limited T_B will become lower as the angle of incidence increases. An example of an object with low emissivity is a smooth water surface, for which ε = 0.4 when observed at normal incidence. Rough soil, on the other hand, has an emissivity close to unity.

An interesting and valuable property of passive microwave radiometry concerns the depth to which ground properties contribute to the observed brightness temperature, i.e.

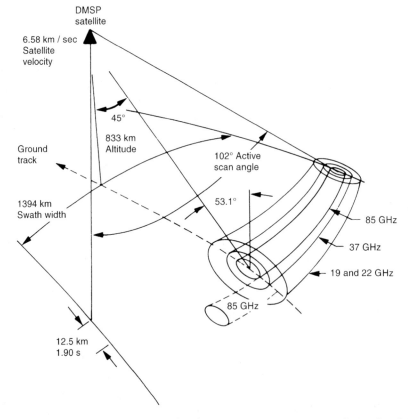

Fig. 4.16 The scan geometry of the DMSP SSM/I. The rotating antenna sweeps the surface in two alternating modes: one in which all four frequencies are recorded; and another in which only 85 GHz data are recorded. The use of a single antenna results in different ground resolutions for each frequency. (Courtesy, Hughes.)

the *penetration depth*. The penetration depth depends on the wavelength used and the dielectric properties of the material. Typical values of the penetration depth are 20 wavelengths for asphalt and sand and only 0.5–0.1 wavelengths for water. Thus wet materials provide little penetration, whereas dry substances, e.g. desert sands, may show substantial (approximately 1 m) penetration. The wavelength dependence of the penetration depth makes profiling of the surface layer possible by using multifrequency radiometry.

Another important parameter characterizing the radiation properties of an object is the *polarization*, i.e. the distribution of the electrical field in the plane normal to the propagation direction. As a rule a radiometer is only sensitive to the field in one direction. Black-body radiation is completely unpolarized, but the emission of many natural features shows pronounced polarization effects which can be useful for identifying the nature of the feature.

In consequence of the above, the measured radiation of a surface feature depends on the following characteristics of the microwave system:

1. the polarization direction of the radiometer;
2. its observation angle related to the surface plane of the object;
3. the frequencies (wavelengths) of bands used.

The radiation received by the microwave system is also influenced by characteristics of the material that is being sensed. The most important of these characteristics are:

1. the electrical and thermal properties of the material;
2. the surface roughness and size of the object;
3. the temperature and its distribution in the body.

Notwithstanding such problems and complications, passive microwave systems can tell us much about the environment that cannot be deduced from other types of passive remote sensing data. Table 4.7. lists some of the uses to

which SSM/I data are now regularly being put. This confirms the great flexibility of such systems – whose future use will be enhanced greatly as and when new radiation collection methods are developed enabling the spatial resolutions of the data to be improved significantly.

Indeed, the chief limitation in the use of passive microwave sensing from space for some environmental applications is the low resolution of the system from orbital altitudes. This is perhaps the chief area in which progress has yet to be made in the evolution of microwave remote sensing.

4.4 ACTIVE SENSORS OUTSIDE THE VISIBLE WAVELENGTHS

4.4.1 MICROWAVE RADAR

The microwave systems discussed so far have been *passive* systems – that is, those receiving natural emissions. However, many *active* sensing systems have been devised from which waves are propagated from bounce-off features in the environment, and are recorded on their return. This is the essence of the radar (radio direction and ranging) system. This was first operated in the VHF radio band (30–300 MHz) but later improvements led to the dominant use of microwave radar (300 MHz to 100 GHz).

Airborne imaging radars fall into two general categories, namely *real aperture* radar (RAR) and *synthetic aperture* radar (SAR). Real aperture radars are commonly associated with side-looking airborne radar (SLAR), an airborne sensor for displaying the back-scatter characteristics of the Earth's surface in the form of a strip image of a selected area. Ground-based radars will be described in Chapter 12.

High-resolution RARs involve a fixed antenna, whose length is conditioned by the size of the aircraft or spacecraft and its effect on the flight performance of the craft. Real aperture radar requires a high power output:

hence RARs are sometimes referred to as 'brute force' systems. Their requirements are difficult to satisfy in satellite systems.

Images of landscape derived from side-looking airborne radar resemble air photographs with low-angle sun illumination, in that shadow effects are produced (Fig. 4.17). This enhances the impression of morphological relief in the imagery but also leads to some obscure areas where the radar shadows occur. However, the geometry of radar pictures is entirely different from that of conventional air photographs. This is because radar is essentially a *distance ranging* device which consequently produces lateral distortion of elevated objects, as shown in Fig. 4.18(a).

The basic elements of a real aperture side-looking airborne radar system (SLAR) are shown in Fig. 4.18(b). A long aerial mounted on the airborne platforms scans the terrain by means of a radar beam (pulse). The radar echoes are recorded by a receiver and processor so that a cathode ray tube (CRT) glows in response to the strength of each echo. The resultant CRT display is then recorded on the film. These radars can be subdivided into six variants:

1. *Monofrequency.* Monopolarized, transmitting a radar carrier polarized in one plane, usually horizontal.
2. *Multipolarized.* This normally transmits a horizontally polarized electromagnetic wave but can receive horizontally and vertically polarized waves separately.
3. *Circular polarized.* This transmits a circularly polarized wave and can receive right-hand or left-hand circularly polarized ground reflections.
4. *Multifrequency.* This transmits two or more modulated radar carriers at widely differing wavelengths to show up differences in penetration and scattering from various surfaces.
5. *Panchromatic.* This transmits a broad-band carrier in order to minimize diffraction effects (speckling).
6. *Polypanchromatic.* A multifrequency panchromatic radar.

In most real aperture SLAR systems the

Table 4.7 Primary environmental products prepared from SSM/I-1 data from launch by the US Navy (Courtesy, US Naval Research Laboratory, 1987)

Parameter	Geometric resolution (km)	Range of values	Quantization levels	Absolute accuracy
Ocean surface wind speed	25	3–25	1	± 2 m s^{-1}
Ice				
Area covered	25	0–100	5	$\pm 12\%$
Age	50	1st year, multiyear	1 year, >2 years	None
Edge location	25	N/A	N/A	± 12.5 km
Precipitation over land areas	25	0–25	0, 5, 10, 15 20, $\geq$25	± 5 mm h^{-1}
Cloud water	25	0–1	0.05	± 0.1 kg m^{-2}
Integrated water vapour	25	0–80	0.10	± 2.0 kg m^{-2}
Precipitation over water	25	0–25	0, 5, 10, 15, 20, $\geq$25	± 5 mm h^{-1}
Soil moisture	50	0–60	1	None
Land surface temperature	25	180–340	1	None
Snow water content	25	0–50	1	± 3 cm
Surface type	25	12 types	N/A	N/A
Cloud amount	25	0–100	1	$\pm 20\%$

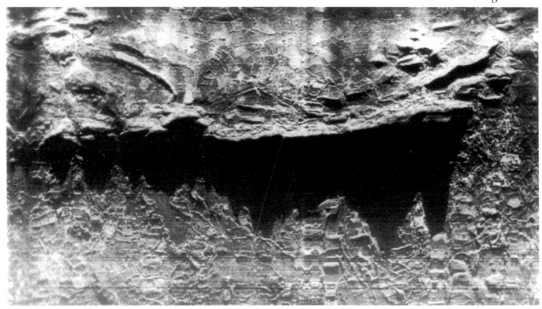

Fig. 4.17 Radar imagery of the Malvern Hills. (Courtesy, Royal Signals and Radar Establishment, Malvern. Crown copyright.)

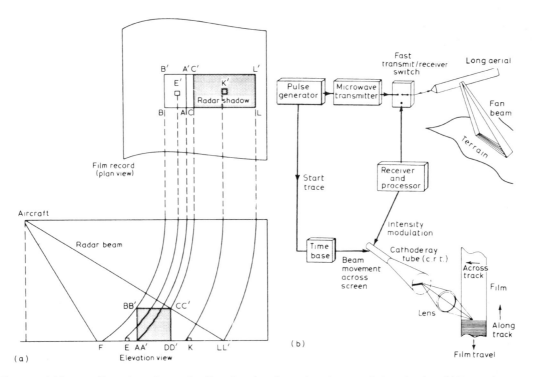

Fig. 4.18 (a) Image distortions due to the directional and ranging characteristics of radar. (b) Basic elements of a real aperture radar system (side-looking airborne radar). (Source: Grant, 1974.)

resolution across the film record is up to 10 times better than that achieved in the along-track direction. The latter can be improved only by increasing the size of the antenna used. There is a limit, however, to the size of the antenna that can be mounted on aircraft and spacecraft.

In fact, further improvements in real aperture resolution became increasingly hampered by limitations in:

1. *millimetric radar technology*, where it becomes difficult to acquire components for engineered radars at wavelengths less than 8 mm.
2. *mechanical engineering tolerances* on antenna manufacture. Even at 8 mm the demanded tolerances are approximately 0.1 mm over a 5 m aperture. Furthermore, these must not only be achieved in manufacture but also maintained in flight in the face of temperature and vibration effects.

So it was that the alternative technique of *aperture synthesis* was developed, whereby the

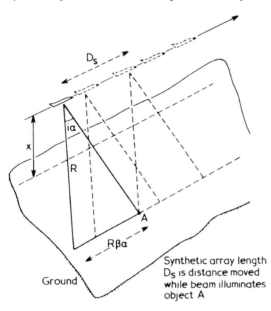

Synthetic array length D_S is distance moved while beam illuminates object A

Ground

Fig. 4.19 Schematic diagram to show elements of a synthetic aperture radar system. (Source: Grant, 1974.)

movement of the aircraft (or spacecraft) along its track can be used to form a long synthetic aerial (Fig. 4.19). This system depends on a particular method of processing the Doppler information associated with the radar returns, and is termed the 'synthetic aperture' system.

4.4.2 SYNTHETIC APERTURE RADAR

The synthetic aperture system is most easily understood by analogy with more familiar optical concepts. It is known that when a plane wave of monochromatic light falls on a narrow slit the wave on the far side of the slit will diverge to form the familiar 'Fresnel zone pattern' for a slit (Fig. 4.20). The Fresnel zones on the screen are alternately dark and light in tone and are caused by wave interference patterns. The light wave amplitude that causes this observed variation can be shown to be both positive and negative. This represents the variation in the phase of the wave fronts at these points relative to the phase of the wave arriving at the zero point, i.e. the centre of the pattern on the screen.

It would be possible to place a film at screen B (Fig. 4.20) and record the complete intensity image. Alternatively the same photograph could be obtained by traversing a narrow slit across the film (Fig. 4.20(a)) to expose it a piece at a time. If the phase of the waves could be added back it would then be possible to reconstruct an image of slit A by shining a monochromatic (laser) light through the film record.

Now if one compares this example with the synthetic aperture system radar, slit A can be regarded as a small part of the ground terrain. Slit C (Fig. 4.20(b)) is the recording system (CRT) for one item of information obtained by the radar receiver. Screen B (Fig. 4.20(b)) is the complete film record obtained from the CRT display for the pulse striking the ground element represented by slit A.

If the image of the element (A) is to be reconstructed it is necessary to record both the phase and the amplitude of the signals received. This is achieved by carrying a very

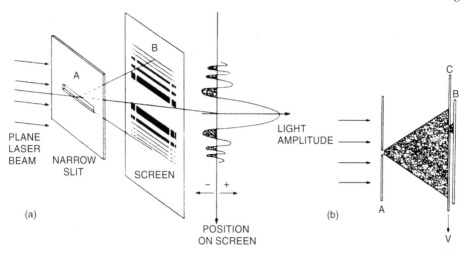

Fig. 4.20 (a) Fresnel zones appearing on screen and diagrammatic appearance of the phases of the waves. (b) Slit (C) moved across the face of the film (B). (Source: Grant, 1974.)

stable reference oscillator in the aircraft or spacecraft and comparing the phase of the return at each position of the sampling slit C with the phase of the stable oscillator. A radar with this type of stable oscillator is known as a 'coherent' radar and it is an essential part of the synthetic radar system.

Any SLAR system acquires an oblique view of the terrain. Subsequent processing provides the interpreter with an image in which the radar geometry has its effect. The images are presented in one of two modes, either 'slant range' or 'ground range'. Slant range and ground range characteristics are shown in Fig. 4.21. *Slant ranged imagery* is related to a timed pulse

to and from a target, whereas *ground range imagery* is heavily dependent on the height of the platform above the ground. If the terrain were absolutely flat a constant factor could be applied, but in practice relief is often uneven and variations in aircraft altitude result in changes in height above ground. The occurrence of relief in the target area introduces a distortion known as *foreshortening*. In such circumstance slopes facing the radar will be shortened relative to their true projected distance (Fig. 4.17 and 4.22). The amount of foreshortening is related to the incidence angle of the radar and so slopes in the near range will be foreshortened more than those in the far range.

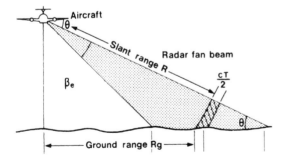

Fig. 4.21 Slant range and ground range. (Source: Trevett, 1986.)

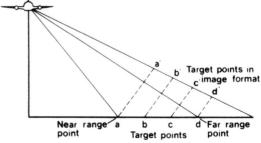

Fig. 4.22 Foreshortening. (Source: Trevett, 1986.)

Thus radar images, even when corrected to give a pseudo-vertical view, have geometric variations across the image dependent on both the incidence angle and the shape of the terrain.

The advantages of side-looking radar compared with conventional aerial photography may be summarized as follows:

1. It has all weather and day/night capability.
2. It is capable of measuring surface roughness and dielectric properties.
3. It provides active illumination, with ability to select wavelength.

The disadvantages of this radar system can be:

1. its poor resolution, although at satellite altitudes the potential resolution of a synthetic aperture system is superior to that of an optical system;
2. its small scale;
3. its image distortion;
4. shadowing in areas of pronounced relief.

The Seasat mission launched in 1978 carried a synthetic aperture radar (SAR) with a frequency of 1.37 GHz providing a ground resolution of 25 m and a swath width of 100 km (section 5.11.2).

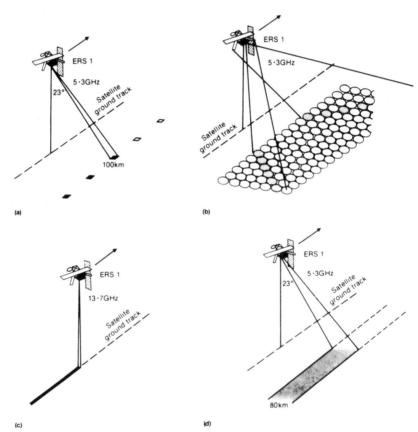

Fig. 4.23 ERS-1: the microwave payload instruments. (a) Wave scatterometer. The SAR is used on a sampling basis as a wave scatterometer obtaining data for the determination of directional wave image spectra. (b) Wind scatterometer, designed to obtain all-weather day and night measurements of surface wind speed and direction. (c) Radar altimeter, which measures the distance between the satellite and the ocean surface and extracts the significant wave height. (d) Synthetic aperture radar mode. (Source: Trevett, 1986.)

4.4.3 SCATTEROMETERS

A particular type of radar system designed to obtain a measure of back-scattered energy for various incident angles of the radar beam is termed the *scatterometer*. The data obtained for each resolution element is presented in a graph of reflected energy versus incident angle. The variation of radar returns can be used by the scientist to determine roughness, texture, and orientation of the terrain. Different surface materials also may be identified.

Scattering signatures have been investigated experimentally for many materials, and theory has been developed for statistical roughness parameters and electromagnetic parameters of various materials. Operational scatterometers are used to gather data along a strip underneath the spacecraft or aircraft. Resolution capability is of the order of 1 m. On Skylab a narrow round beam was used which discretely shifted from the maximum angle to the nadir of the spacecraft as it moved over a given area.

A major new European satellite venture in recent years has been that of the launch of the first two ERS satellites. ERS is designed primarily as a sea and ice observation satellite and for this purpose it carries a radar altimeter and wind and wave scatterometers as well as an imaging radar in C band. The general characteristics of the entire suite of ERS-1 microwave radars are shown in Fig. 4.23. The SAR imaging radar on the Canadian Radarsat, launched in 1995, is broadly similar: its imaging modes are shown in Fig. 4.24.

4.4.4 LIDAR (OR LASER RADAR)

The lidar is an active system similar to a microwave radar, but operating in that part of the spectrum comprising ultraviolet to near-infrared regions. It consists of a laser which emits radiation in pulse or continuous mode through a collimating system, and a second optical system which collects the radiation returned and focuses it on to a detector. Three types of lidar are at present available: an altimeter type, which can plot a terrain profile; a scanning type, which can be used as a mapping instrument; and a third type using spectroscopic techniques, which may be used for mapping air pollutants.

Several techniques evolved in the 1970s for measuring three-dimensional wind velocities with continuous wave Doppler lidar systems. A compact, remotely operable airborne system

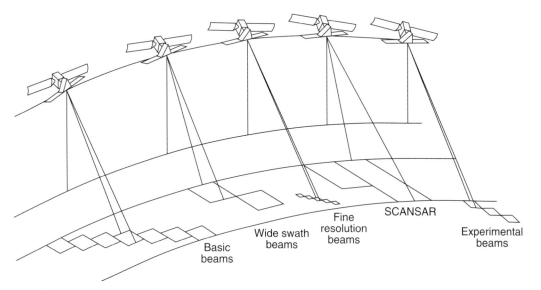

Fig. 4.24 The Radarsat SAR imaging modes. (Courtesy, Radarsat.)

was built at the Royal Signals and Radar Establishment, UK. An outstanding feature has been its long-term reliability and as a result pulsed CO_2 lidars have set the scene for the development of spaceborne lidar wind measuring systems to measure global wind profiles. The ongoing programme of development of the Laser Atmospheric Wind Sounder (LAWS) represents an important application of lidar technology. Another is detailed surface mapping and monitoring from aircraft altitudes, as described in Chapter 17.

5.1 INTRODUCTION

Remote sensing techniques can be applied to data from different types of observation platform. Each platform, mobile or stationary, has its own characteristics. Generally speaking, three types are of interest for remote sensing: *ground*, *airborne* and *spaceborne* observation platforms. A classification of remote sensing platforms based on these three categories is given in Fig. 5.1.

Each type of platform has its own particular advantages and disadvantages. For example, the satellite platform offers great synoptic viewing potential, stability of the platform, and orbital movements of predictable character. Yet the space platform is difficult to reach and instrumental failure may be hard to correct. Airborne platforms offer easier maintenance of sensors, lower operating costs and good accessibility for manned flights. Yet airborne platforms may suffer from instability

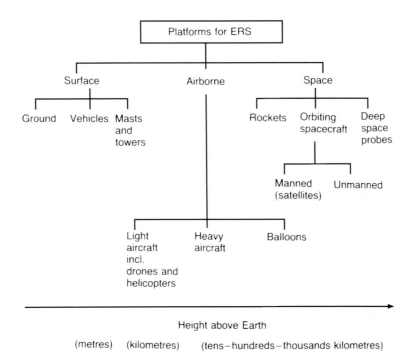

Fig. 5.1 A classification of remote sensing platforms by type and altitude.

and flight lines may be difficult to maintain. Therefore, it is not uncommon for remote sensing monitoring programmes to involve multiplatform experiments operations, as illustrated in Fig. 5.2. The following sections in this chapter discuss each type of platform with these issues in mind.

To make effective use of any remote sensing system, a number of factors should be considered in the planning phase, including:

1. The *time and space scales* of the environmental features that are essential to the fulfilment of the objectives of the programme.
2. The *time of day and/or the season* when the features of concern can best be observed, e.g. related to the phenology of a crop or natural vegetation cover, the seasonal/ diurnal behaviour of animal species, or weather feature development such as thunderstorms (diurnal growth cycles) or hurricanes (seasonally distributed).
3. *The effects of key physical influences* (e.g. solar illumination, surface temperature and surface moisture) on the imagery obtained.
4. Possible *atmospheric attenuation* of the energy collected by the sensor.
5. The *repetition rate of observations* required to resolve the questions it is hoped to solve by remotely sensed means.

5.2 GROUND OBSERVATION PLATFORMS

In many areas of study, a preparatory programme is necessary to provide a good understanding of the basic physics of the

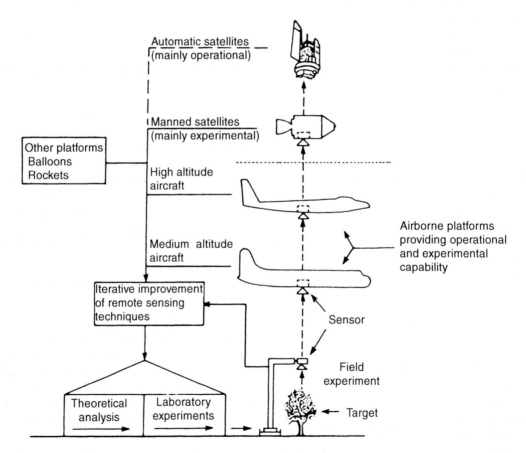

Fig. 5.2 Multiplatform remote sensing operations.

object/sensor interaction. Measurements of physical characteristics, especially the spectral reflectances and emissivities of different natural phenomena, are necessary in order to design and develop sensors. Normally such studies are of two kinds:

1. *Laboratory studies* where soils, plants, building materials, water bodies, etc., are subjected to an external source of radiation and detectors of various kinds are used to measure reflection and emission spectra.
2. *Field investigations* in which the spectral characteristics of surface phenomena or crops are investigated in different atmospheric conditions.

Many studies of spectral response and emissivity are carried out by means of hand-held radiometers. These are, however, limited by such factors as field of view, weight of sensor, periodicity of observations and continuity of data collection. These limitations have prompted the use of various types of ground-based platforms for the more repetitive collection of highly detailed data by remote sensing means.

In field investigations some sensors may actually be located on the ground itself, or on platforms at or very near ground level, e.g. cameras on one hillslope to record overland flow on an opposite hillslope. Amongst mobile ground platforms, the most versatile are the 'Cherry Pickers' (Fig. 5.3). These can be extended to approximately 15 m, and through a full circle. They have been used by various research organizations to carry sensors such as spectral reflectance meters, photographic systems and scanners in the infrared or radar wavelengths. Frequently these platforms are linked to automatic recording apparatus, and mounted on field vehicles not limited to roads.

Portable masts are also popular ground observation platforms. They are available in various forms and are often used to support cameras and sensors for testing. For example, a Land Rover fitted with an extending mast may be useful in conjunction with airborne

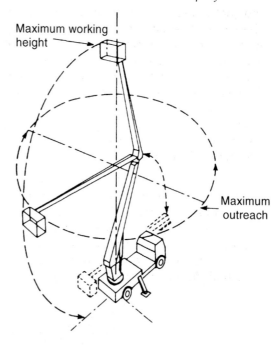

Fig. 5.3 Mobile hydraulic platform of the 'Cherry Picker' type.

studies (Fig. 5.4(a)). However, such masts can be very unstable in windy conditions. Towers, either fixed or temporary (Fig. 5.4(b)), are also available. These offer greater rigidity than masts, but may require more time to erect. Where heavy instruments, e.g. radar systems, are undergoing tests it may be more effective and economic to place towers on a wheeled platform so that they can be moved on rails.

5.3 AIRBORNE OBSERVATION PLATFORMS

5.3.1 BALLOON PLATFORMS

Balloon platforms for remote sensing instruments can be considered under two headings: *tethered* and *free flying*. Tethered balloons were used for remote sensing observations (air photography) as long ago as the American Civil War (1862) and very recently for nature conservancy studies on beaches in Britain. Modern studies, however, are often carried out by free-flying balloons, which can offer

(a)

reasonably stable platforms up to very considerable altitudes. For example, the balloons developed by the Société Européene de Propulsion can carry a photographic nacelle (Fig. 5.5) to altitudes of 30 km. This nacelle consists of a rigid circular base plate supporting the whole equipment to which is nested a tight casing, externally protected by an insulating and shock-proof coating. It is roll-stabilized and the inner temperature is kept at 20 K and at ground atmospheric pressure. Standard equipment includes two cameras, a multi-spectral photometer, power supply units and remote control apparatus. Its return to Earth is achieved by parachute, after remote controlled rupture of the carrying balloon.

5.3.2 POWERED AIRCRAFT

Commercially oriented remote sensing from aircraft has been carried out for more than 60

(b)

Fig. 5.4 (a) An extending mast fitted to a Land Rover. The mast can be extended to 10 m height and is equipped with camera and photodiode sensors for studies of crop reflectance. (Courtesy, University of Bristol.) (b) Tower for supporting sensors. (Courtesy, Hawker Siddeley Dynamics Ltd, Hatfield, Herts.)

(a)

(b)

Fig. 5.5 (a) Balloon nacelle before flight. (b) Sensor equipment inside balloon nacelle. Photometer (black) and cameras in base plate. (Courtesy, Société Européene de Propulsion.)

years using fixed-wing aircraft. Increasingly in recent years helicopters have been used widely for such purposes also. Historically the fixed-wing aircraft have been used mainly for obtaining stereoscopic black and white photography for subsequent interpretation and photogrammetric mapping (Chapter 7). Many resource development and planning programmes use aerial photography from altitudes of approximately 1500–3000 m. These missions at low or medium altitudes are appropriate for surveys of local or limited regional interest. Although they are used widely for periodic surveys it must be recognized that flights at specified times at short intervals or for defined repetition rates over long periods are rarely achieved. Nevertheless, aircraft platforms offer an economic method of testing sensors under development.

In recent years the rise of aircraft as platforms for remote sensing has broadened greatly, involving not only a range of photographic cameras but also many different scanners in the infrared, radar and microwave wavelengths.

In order to obtain a greater area coverage, and to test atmospheric effects on sensors, special high altitude aircraft have been used as platforms. These have usually been developed from military reconnaissance aircraft, such as the American U2. The altitudes attained by such aircraft are up to about 15 km, from which ground coverage of 50–400 km^2 per frame can be obtained.

In some cases the use of 'normal' aircraft may be deemed too expensive, or it may be that there is a demand for flights at very low levels and at low speeds. In such circumstances the

Fig. 5.6 (a) Remotely piloted vehicle (RPV) (drone), which is launched from a trailer. (Courtesy, AERO Electronics (AEL) Ltd, Horley, Surrey.)

microlight (ultralight) aircraft which have been developed in recent years are promising platforms. For example, a microlight equipped with two radiometers, vidicon camera, portable stereo videorecorder and a laser range finder has been used for low-level flights at 5–10 m altitude over specific test sites in the UK.

Today helicopters are also widely used for environmental studies. In studies of inaccessible areas (e.g. swamps and bogs) they can carry out quadrat-like studies speedily and economically. Helicopters are now capable of carrying quite heavy loads over substantial distances and are used widely as photographic camera, vidicon camera and thermal heat sensor platforms.

A new type of airborne platform has emerged recently as a result of military requirements for target systems. This is variously termed the 'remotely piloted vehicle' (RPV), the 'unmanned air vehicle' (UAV) or more commonly the 'drone'. A drone is a radio-controlled miniature aircraft. Since it is controlled from the ground its relatively short control range of 5 km is dictated by the visual range obtainable. However, aerial photography of local, inaccessible areas is highly practicable. For example, surveys of oil pipelines, fence lines, power transmission lines or forest areas are possible using an underwing robot camera which is triggered by remote control signal from the ground. Whilst fixed-wing drones were most popular a decade or two ago (Fig. 5.6(a)), rotary-wing (mini-helicopter) drones are now considered more flexible (Fig. 5.6(b) and (c)).

5.4 HIGH-ALTITUDE SOUNDING ROCKETS

In the early 1970s some interesting experiments in the use of sensors from space altitudes were carried out using modified

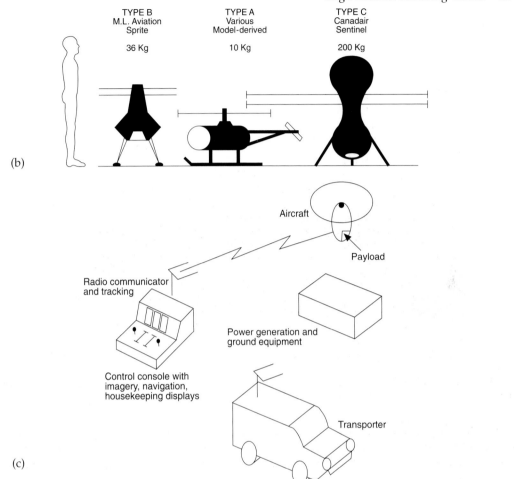

Fig. 5.6 (b) Helicopter types of unmanned air vehicles. (c) Elements of a vertical takeoff and landing (VTOL) drone. These are similar to fixed-wing systems but without launch and recovery equipment. All may be carried in one transporter, as the aircraft is more compact. (Source: Austin, 1994.)

sounding rockets as the platforms. One limitation upon the use of such a system was that there was a need to ensure that the descending rocket did not cause damage. As a result its use was limited to sparsely populated areas, e.g. in surveys over Australia and Argentina.

The Skylark Earth resources rocket (Fig. 5.7) was an example of such a platform. The rocket was fired from a mobile launcher to altitudes between 90 and 400 km. During flight the rocket motor and payload separated and the sensors were held in a stable attitude by an automatic control system. By stepping the payload around six times, the sensors scanned a 360° field of view. The payload and the spent motor returned to ground slowly by parachute, enabling speedy and safe recovery of the photographic records. The sensor payload consisted of two cameras, normally of the Hasselblad 500EL/70 mm type.

Unfortunately the imagery was obtained from a range of different altitudes and look-angles, posing problems for subsequent use of the data. Improvements in the spatial

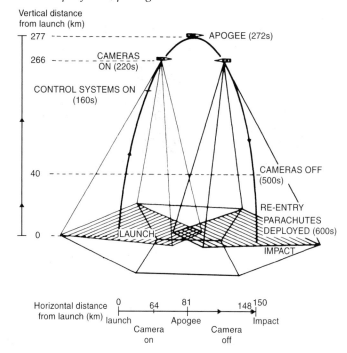

Vertical distance
from launch (km)

277 APOGEE (272s)

266 CAMERAS ON (220s)

CONTROL SYSTEMS ON (160s)

40 CAMERAS OFF (500s)

RE-ENTRY PARACHUTES DEPLOYED (600s)

LAUNCH

IMPACT

0

Horizontal distance from launch (km)   0   64   81   148 150
launch   Apogee   Impact
Camera on   Camera off

Fig. 5.7 Schematic diagram of Skylark rocket trajectory and photographic coverage. (Source: Savigear *et al.*, 1974.)

resolutions of data obtained from satellites seem likely to have closed the Earth resources rocket chapter, though fitted with a digital scanning device such a cheap platform could have a brighter future owing to the rapid advances that have been made with the handling of digital rather than photographic image data.

5.5 SATELLITE PLATFORMS

5.5.1 ORBITAL CONSIDERATIONS

It is now generally recognized that most environmental monitoring can be achieved best by a *combination of missions* embracing a variety of sensors and sensor packages. However, the satellite has quickly become the key platform in many programmes and projects. As well as providing the broadest views of regional relationships, satellite platforms can be placed into a variety of orbits that ensure repeated coverage of the whole of the Earth's surface. The interval between satellite observations at a particular location can, to some extent, be

selected by choice of an appropriate orbit. Furthermore, such orbits can be so defined that observations can be made under comparable or identical conditions of illumination over comparatively long periods. Some description and explanation of orbital configurations is appropriate before considering details of the platforms and sensor packages themselves.

The orbital relationships of satellites are affected by many factors but one may note first that there is a basic relationship between the *heights of satellite orbits* and their *lives before re-entry* to the Earth's atmosphere. Satellites in low orbits are affected much more by upper atmospheric drag and resulting increases in the effects of gravitational pull (Table 5.1).

Thus there are lower limits below which Earth observation satellites will not normally be operated. Since the design lives of civilian payloads are usually within the span from 1 to 5 years, a minimum operational altitude of about 450 km is indicated. Some classes of military satellites operate much lower than this, providing very high resolution data on quickly recovered aerofilm. Actual orbital

Table 5.1 Relationships between average altitudes of satellites in near-circular orbits, and their expected lives before re-entry

Height above Earth surface (km)	Life before re-entry
250	12 days
500	10 years
600	50 years
1 000	1000 years
10 000	Indefinite

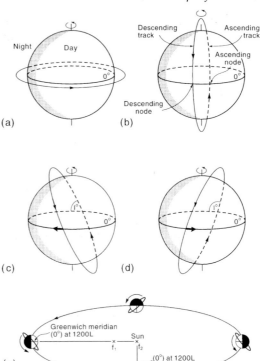

Fig. 5.8 Orbital characteristics of polar-orbiting satellites, and the Earth as a satellite of the Sun. (a) Equatorial orbit. (b) True polar orbit. (c) Prograde oblique orbit (launched eastwards, regresses westwards). (d) Retrograde oblique orbit (launched westwards, precesses eastwards). (e) Perpendicular from Sun to Earth, precesses eastwards throughout the year.

altitudes of unmanned remote sensing satellites are characteristically well above that level; whilst most operate between 800 and 1500 km, some are at approximately 36000 km altitude. The first of these groups is composed mostly of *polar or near-polar orbiting* satellites occupying so-called sun-synchronous orbits (crossing the Equator at the same sun time each day); the second group are mostly *geostationary* satellites over preselected points on the Equator.

Although all satellite orbits are elliptical, for most Earth observation purposes it is desirable to achieve as nearly a circular orbit as possible. However, this is not always the case. Consider two satellites launched on the same day, 27 June 1979. Of these, NOAA-6 was placed into a nearly circular orbit. The apogee (top) of the NOAA-6 orbit was 826 km, and the perigee (bottom) was 810 km. A Russian satellite launched on the same day was Cosmos 1109. Its orbit was highly elliptical with an apogee of 40 058 km and a perigee of 613 km: such orbits may actually be preferred for satellites whose viewing missions differ from one region of the world to another. (See also Fig. 5.9.)

5.5.2 LOW-LEVEL SATELLITES

Low-level (800–1500 km) Earth observation satellites ('LEOs') encircle the Earth whilst the Earth itself revolves on its polar axis within their orbits. These satellites fall into three broad groups on the basis of detailed orbital differences:

1. *Equatorial-orbiting satellites*, whose orbits are

wholly within the plane of the Equator (Fig. 5.8(a)).

2. *Polar-orbiting satellites*, whose orbits are in the plane of the Earth's polar axis; each successive orbit crosses the Equator at a different sun time (Fig. 5.8(b)). True polar orbits are preferred for (rare) missions whose aim is to view longitudinal zones under the full range of illumination conditions.

3. *Oblique-orbiting (near-polar orbit) satellites* (of which those in near-polar orbits comprise by far the largest group), whose orbital planes cross the plane of the Equator at an angle other than 90°. Because the range of

possibilities here is much greater than in (a) and (b), further discussion of oblique orbits is required (Fig. 5.8(c) and (d)).

Oblique-orbiting satellites may be launched eastwards into direct, or *prograde*, orbits (so called because the movement of such satellites is in the same direction as the rotation of the Earth), or westwards into *retrograde* orbits. The inclination of any orbit is specified in terms of the angle between its ascending track and the Equator, i.e. at the ascending node. Because the Earth is not a perfect sphere it exercises a gyroscopic influence on satellites in oblique orbits such that those in prograde orbits (Fig. 5.8(c)) *regress* whilst retrograde orbits *advance* or *precess* (Fig. 5.8(d)) with respect to the planes of their initial orbits.

Figure 5.8(e) shows that a perpendicular from the Sun to the Earth also advances around the globe in the direction of the Earth's rotation on its polar axis, because of the behaviour of the Earth itself as a satellite travelling around the Sun. It is clear that, if the rate of precession of an orbit is geared to the orbiting of the Earth around the Sun, a special type of Earth satellite orbit can be achieved: this is the *sun-synchronous* configuration (a particular retrograde oblique orbit) which ensures that satellites view the same point on the surface at the same local time at predetermined intervals (twice each day with near-polar orbiting weather satellites, typically once every several days with Earth resources satellites). Thus the altitude of the Sun in the sky is approximately the same within similar seasons at every crossing of the same parallel. However, it should be noted that differences in the elevation of the Sun affect each local area through the annual cycle of the seasons as a result of the apparent motion of the Sun relative to the Earth.

It is in this context that the observation of the surface from a satellite must be seen to be a function not only of the satellite altitude and orbital configuration, but also of the field of view (FOV) of its sensor system, i.e. the angle through which a sensor can collect radiation from its target.

One very desirable sun-synchronous orbit is a retrograde orbit inclined at 99.1° to the Equator, at an altitude of 1100 km. This orbit is completed once every 100 min. From this information the number of orbits that will be completed in 24 h can be calculated as approximately (14.5), as also can the breadth of the FOV required to ensure complete global imaging cover even in equatorial latitudes. In the case of NOAA satellites, once-daily coverage is achieved in the visible by a scanning radiometer with a viewing angle of 112°; twice-daily coverage is achieved in the infrared by the same sensor system because heat radiation is emitted from both the daytime (descending track) and night-time (ascending track) hemispheres. In the case of Landsat, narrower swaths of imagery have been obtained from scanning radiometers, such as the Multispectral Scanner (MSS) of Landsat 3, which had a viewing angle of only 11.6°, and the satellite occupied a lower orbit, at about 900 km. In such ways more detailed target information is acquired, but at the expense of breadth and frequency of coverage. *Such trade-offs are fundamental in satellite remote sensing*, for communication links between satellites and ground stations constrain the volumes of data that can be transferred from the one to the other whilst they are intervisible.

5.5.3 HIGH-LEVEL SATELLITES

Whilst the sun-synchronous orbit is the most popular low orbit, *geosynchronous* orbits are virtually the only useful high orbits. In these the satellites ('GEOs') precess around the Earth at rates related to the rotation of the Earth on its polar axis. The most useful single orbit is at approximately 36 000 km, at which altitude a prograde satellite orbiting in the plane of the Equator 'hovers', i.e. it is *apparently 'fixed' above a given point on the surface*. This special type of geosynchronous orbit is known as *geostationary*. It is exploited by many communications satellites, and some Earth observation satellites, especially within the

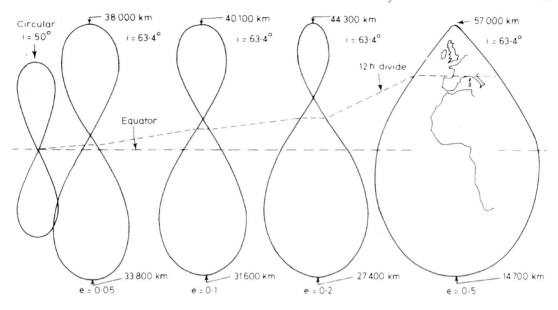

Fig. 5.9 Ground tracks of inclined geosynchronous orbits to exemplify possible variations of eccentricity, inclination and orientation. Associated variations would result in imaging resolutions (possibly advantageous) but also in image geometries (generally disadvantageous). (Source: Velten, 1976.)

programme of the World Weather Watch (WWW) of the World Meteorological Organization (WMO) (Chapter 10). These have included ATS, SMS, GOES, Meteosat, Insat and GMS satellites. The key advantage of the geostationary orbit is that it permits satellites to view the visible disc (limited by the curvature of the Earth) *much more frequently* than any low orbit satellite can do. Other geosynchronous possibilities exist but have yet to be exploited regularly for Earth observation missions. Figure 5.9 exemplifies some of these, drawn from design studies carried out by ESA.

5.6 THE EVOLUTION OF EARTH RESOURCES SATELLITES

The first photographs of the Earth from space were taken from the NASA spacecraft Mercury, Gemini and Apollo in the 1960s. The interest shown in the 70 mm format colour photographs first taken from a Mercury satellite in 1961 was to grow rapidly as the relief and geological features of many unmapped areas of the world were seen in detail for the first time.

Subsequently some 2400 photographs were taken in colour and infrared colour from the Gemini spacecraft in the mid-1960s. These pictures created considerable interest, and the Apollo mission which followed built upon this, eventually leading to the acquisition of multiband photography using a multicamera array. It was this imagery that laid the foundations for the subsequent series of Landsat sensor packages. NASA's Goddard Space Flight Center (GSFC) began a conceptual study of Earth resources satellites in 1967 as part of this evolutionary stage, and the Landsat family of satellites was the result. Today, the key operational criteria for Earth resources satellites are *high resolution sensors* (surface resolution of tens and hundreds of metres) but also *rather long-repeat imaging cycles* (several or many days).

respectively. The main objectives of the HCMM programme were to conduct research into the feasibility of using day/night thermal infrared remote sensing data for the following:

1. rock type discrimination and mineral resource location;
2. measurement of plant canopy temperature at frequent intervals to determine evapo-transpiration and water stress;
3. measurement and monitoring of soil mois-ture change;
4. mapping natural and man-made thermal effluents;
5. detection of thermal gradients in water bodies;
6. mapping and monitoring snow fields for water run-off prediction;
7. measuring the effects of urban heat centres;
8. monitoring marine oil pollution.

5.11.2 SEASAT

This satellite, specially developed for ocean monitoring, completed 106 days of successful operation before a sudden failure in October 1978. Although Seasat was designed primarily for oceanographic mission objectives carrying a radar altimeter, radar scatterometer, and microwave imaging radiometer, its SAR data were possibly of the greatest environmental interest, being used to study a variety of both sea and land phenomena. High-quality SAR images enabled the following to be carried out:

1. oceanographic monitoring;
2. polar ice mapping;
3. geological mapping to detect lineaments, folds, faults, fractures;
4. drainage network analysis;
5. crop type and growth stage;
6. vegetation mapping and soil studies.

The most important feature of Seasat for environmental monitoring and management was that it demonstrated comprehensively for the first time the use of an active microwave sensor, which was an all-weather capability. There is little doubt that had Seasat not failed

the data would have found wide applications. It is interesting to note that the Japanese MOS (section 5.6), the Earth Resources Satellite launched by the European Space Agency (namely ERS-1 and 2), and the Canadian Radarsat have similar radars as their prime sensors.

5.11.3 THE NIMBUS-7 COASTAL ZONE COLOR SCANNER (CZCS)

The CZCS was launched on Nimbus-7 in 1978, and continued to provide good data until the late 1980s. It was the first satellite sensor dedicated to monitoring coastal zone and ocean environments. This was a five-channel instrument resolving at *c.* 800 m, with four visible waveband channels, plus one in the infrared. Its repeat cycle was nominally 6 days. In particular the main objectives of CZCS included:

1. mapping and measurements of suspended materials and other phenomena over large areas of water;
2. improving scientific knowledge of marine ecosystems;
3. assessment of existing fisheries and potential fishing grounds;
4. experimenting with real-time data acquisition and the rapid production of sea parameter maps;
5. defining requirements for future ocean monitoring instruments.

The new SeaWiFS (Sea-viewing Wide Field-of-view Sensor) satellite launched by the USA in 1997 was designed, and has proved, to be a worthy and much needed successor to the CZCS for ocean colour monitoring and analysis (section 5.12.2).

5.12 NEW GENERATION SATELLITES

5.12.1 THE EUROPEAN REMOTE SENSING SATELLITES (ERS)

In order to achieve full benefit from remote sensing it is necessary to look towards all-

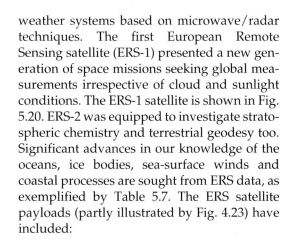

Fig. 5.20 The ERS-1 satellite. (Courtesy, ESA.)

weather systems based on microwave/radar techniques. The first European Remote Sensing satellite (ERS-1) presented a new generation of space missions seeking global measurements irrespective of cloud and sunlight conditions. The ERS-1 satellite is shown in Fig. 5.20. ERS-2 was equipped to investigate stratospheric chemistry and terrestrial geodesy too. Significant advances in our knowledge of the oceans, ice bodies, sea-surface winds and coastal processes are sought from ERS data, as exemplified by Table 5.7. The ERS satellite payloads (partly illustrated by Fig. 4.23) have included:

1. an Active Microwave Instrument (AMI), which operates in three different modes;
2. a Radar Altimeter, which provides accurate measurements of sea-surface elevation, significant wave heights, sea-surface wind speeds and various ice parameters;
3. an Along-track Scanning Radiometer (ATSR) and Microwave Sounder combining infrared and microwave sensors for the measurement of sea-surface temperature, cloud-top temperature, cloud cover, and atmospheric water vapour (the ATSR on ERS-2 investigates four reflected channels, in the visible and infrared);
4. Precise Range and Range-rate Equipment (PRARE) for the accurate determination of satellite position and orbit characteristics and for geodetic 'fixing' of ground stations;
5. Laser Retro-Reflector (LRR) for the measurement of satellite position and orbit using laser ranging stations on the ground.

separated intervals of time, so there are errors attached to such observations which need to be assessed. For example, if airborne or spaceborne sensor data are being used to identify agricultural crops, it is necessary to know the condition of a sample population of fields before assessing the accuracy of interpretation of the remote sensing data set itself. It is now recognized that there are many factors which may introduce *errors* into such field surveys. These include boundary/locational problems, ambiguity in crop classification, inconsistency between surveyors and time lapses between the surveys and the acquisition of the related remotely sensed data. Further, the range of observations and so-called surfaces observed or observable is potentially very broad. Sometimes it is difficult even to define the measured 'surface' very precisely; for example, because depth-related factors may be involved, e.g. wind-speed, solar radiation, rainfall, atmospheric humidity and cloud cover. Measurements of such parameters may be made close to the ground (microclimatic observations), whilst others may be obtained from standard meteorological screens (mesoclimatic observations). Others may be recorded at considerable heights in the atmosphere by radiosonde techniques.

Similarly a wide variety of methods are employed for measuring *in situ* conditions in water bodies. Oceanographic observations may include measurements of sea temperature, salinity, wave motion (height and wavelength) and biological content as well as meteorological conditions. These measurements are made from a variety of platforms, including weather ships, coastal protection vessels, automatic data collection buoys and coastguard stations, again with no general standardization of observation method.

In estuaries and rivers additional factors, such as suspended sediment load, biological oxygen demand, water reaction and pollution, are often recorded. These observations may be made from boats or by automatic samplers used from the banks.

It should be noted, too, that large-area atmospheric, hydrologic and oceanographic studies commonly depend on data from extant stations or station networks. However, special efforts to collect *in situ* data in these and other contexts are often necessary and *great care is needed to ensure that the two resulting types of data sets are as compatible and complementary as possible.*

In this chapter we shall draw examples mainly from ground and water surface studies, but the general problems of data collection, such as sampling and selection of sites for observations, are relevant to other types of surfaces also. The data transmission methods are always essentially the same in each case.

Finally, a word about the concept of 'conventional observations'. Sometimes this is taken to be a synonym for *in situ* data. However, the concept should be used with care, for some remote sensing methods (e.g. weather radar methods in meteorology) have become so well-established as to be viewed by some at least as being 'conventional' *vis-à-vis* satellite techniques. Therefore, like some other terms and concepts discussed in this section, the concepts of 'conventional data' or 'conventional observations' are generally best avoided.

6.2 SELECTION OF GROUND DATA SITES

The location of areas for ground data collection in support of particular aircraft and/or satellite remote sensing 'campaigns' may be decided on the basis of a number of criteria. These include study objectives, the sample sizes satisfactory for statistical purposes, repeatability and continuity of the experimental study, access to the study area, availability of existing *in situ* data for the area, trained personnel, equipment resources and the orbit characteristics of the space platform.

The *shape* of the ground data collection area is dependent on statistical requirements and speed of access. The allocation of sample plots in a data collection area is made more statistically efficient if the area can be stratified into relatively homogeneous areas. For stratification to be useful, strata boundaries should separate areas where within-class variance is less than between-class variance. The number of survey plots can then be estimated by the method of proportional allocation. Thus, the shapes of homogeneous areas may influence the shapes of ground survey areas. In addition, the sampling technique used in the ground observations (grid, area, line) may influence the shape of the area; for example; line sampling along existing road networks may be preferred to block area sampling.

The *size* of the ground data collection area will be affected by study objectives, statistical considerations, scale of the ground phenomena, angle of view of the sensors and time factors. Where sensor testing is the objective, it is desirable to select the smallest ground area that allows detailed ground monitoring by equipment and personnel over a wide range of ground/atmospheric conditions. Many small sites may be necessary if it is desired to check on the validity of a spectral signature for a particular surface condition. An example of such a site is that used by one of the authors at Long Ashton near Bristol (Fig. 6.1). The total size of the site is 120 ha, within which smaller areas of approximately 2 ha were repeatedly monitored.

Where relationships are sought between the sensor response and particular surface conditions (e.g. a particular crop such as grass) it is necessary to obtain sufficient samples of the crop (generally more than 30) to allow statistical tests to be carried out. The size of area that will provide sufficient samples must be determined.

Time constraints also affect the organization of ground data collection activities, through:

1. the quantity of data required;
2. the resources available for collection
3. the rate of change in ground environmental conditions.

For example, where evaporation rates are high it may be necessary to monitor changes in soil moisture content frequently (several times a day), whereas in conditions of low evaporation loss the soil moisture change may be slow and fewer observations are required in a given time.

The prime objective of ground data collection is to provide a contemporaneous record of ground conditions at the time of remote sensing. In practice it is difficult to obtain synchronous data for more than a small area or selected sample sites. The aim, however, is to obtain sample ground truth data within a short time of the acquisition of remotely sensed data.

In planning ground data collection, special attention should be given to the rate of change of the variables to be observed. These variables can be categorized as *transient* or *non-transient*.

Fig. 6.1 Ground truth site, Long Ashton, N. Somerset. Note the soil moisture measuring equipment in the foreground, including neutron moisture probe and tensiometers. (Photo: L.F. Curtis.)

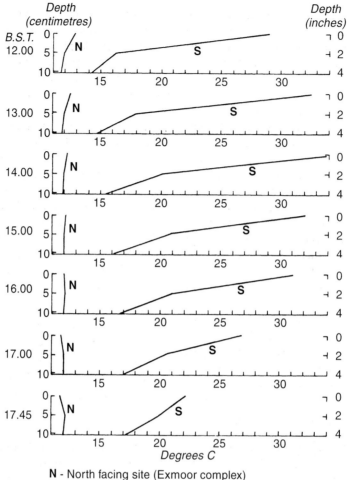

N - North facing site (Exmoor complex)
S - South facing site (Exmoor complex)

Fig. 6.2 Contrasts in soil temperatures on north and south facing slopes on Exmoor, England.

Data recording of transient features (e.g. crop stage, leaf cover, wind speed, surface moisture) must be near synchronous. Recording of non-transient features (e.g. slope, aspect, soil texture) can be carried out prior to, or after, the sensing mission.

As an example of rate change in a transient surface condition, one may note that data for soil temperature variation on slopes of an Exmoor Valley (Fig. 6.2) show that frequent observations are necessary on south-facing slopes, but the rate of change on north-facing slopes is much less.

The nature of the ground data required varies according to the type of investigation being made. Geological surveys often demand that rock and soil samples be taken for analysis and description (Fig. 6.3). Hydrological studies require a range of information, including stream gauging, suspended sediment contents, water-table measurements and local climatic data. In studies of soil conditions the ground truth data normally include records of soil phase (or soil series), soil moisture, soil temperature, soil mixture, structure, stoniness, organic matter content, soil colour and bulk density.

Some data can be obtained only a short time before the remote sensing mission takes place because the phenomena under investigation are continually changing. For example, in crop studies it is necessary to observe transient features such as the type, stage height and colour of crop, disease types, weed species, effects of husbandry (ploughed, harrowed, drilled, rolled, wheeling marks), the grazing method, number of livestock and per cent crop cover.

Other types of data are more permanent and features such as the morphology of the terrain can normally be recorded by field survey and analysis of contour maps before the sensing missions take place. The morphology of the ground over which the mission is carried out is an important element in respect of data interpretation. Gradient, slope form and aspect often have significant effects on sensor data. This is particularly the case where radar data are being used. Terrain classification and evaluation techniques are now well developed and they can be used for recording site morphology at various levels of detail.

6.3 COLLATERAL DATA FOR CROP STUDIES

In this section selected aspects of supplementary or collateral data collection for crop monitoring will be reviewed. It must be

Fig. 6.3 Field examination of rock and soil samples in relation to air survey. Adequate labelling and field description are essential elements of camp work. (Courtesy, Hunting Technical Services Ltd, Elstree.)

emphasized, however, that each combination of sensor(s) and user (applications) requires careful consideration before any particular mission takes place: frequently, the nature of the data to be collected will be affected by the geographic location of the area of investigation.

The range of collateral data required varies according to the farming region and its soil, relief and climatic conditions. Furthermore, the cost of acquisition of remote sensing imagery often determines that it be used for more than one purpose, e.g. in the agricultural context not only for crop recognition, but also for soil drainage mapping and land quality evaluation. Generally, four categories of data are required for multipurpose land use studies in rural areas (Fig. 6.4):

1. site morphology;
2. crop/vegetation cover characteristics;
3. cultivation/husbandry features;
4. soil surface conditions.

The range of data to be collected should also be related to the organizational structure and personnel resources. A summary table (Table 6.1) gives manpower requirements for ground data collection in respect of integrated studies in three English counties. The prime objective of ground data is to provide a contemporaneous record of ground conditions at the time of remote sensing. In practice it is rarely possible to obtain detailed synchronous agricultural data for more than a small area or selected sample sites. For example, data for spring barley in Nottinghamshire showed that mean percentage leaf cover increased from 18 to 40 in a period of 8–10 days in the first half of May. These changes are of sufficient magnitude to necessitate repetition of ground data collection since the proportion of bare soil exposed beneath a growing crop will have a major effect on image response. Quantitative observation of soil exposure using quadrat sample methods are time consuming. A single observer measuring crop cover using a 50 × 50 cm, 100-point quadrat for 500 observations

DATE . . . .   LAND USE * . . . .   FIELD REF . . . .

CROP CONDITIONS

SOIL CONDITIONS

Stage   *

Height: average
            range
            pattern   *
            extent
            comment

Colour: general   *
            pattern   *
            extent   *
            comment

Disease: type
            extent   *
            comment

Weeds: species
            density   *
            comment

Husbandry
            Ploughed
            Harrowed
            Drilled
            Rolled
            Wheelings

            Grazing
                method
            Livestock

% Soil exposed

Surface   general
colour: pattern   *
            extent   *
            comment

Roughness: furrowed
            normal tilth
            cloddy
            panned

Surface   abundance
stones: %
            size
            type

Surface
moisture   *

*Site
Morphology:*
            Gradient   *
            Slope type *

            Microrelief
                type
                extent *
            Field   *
                Boundary

* Codes available (e.g. extent --
  1: < 5%; 2: 5-50%; 3: > 50%)

General Comments

Fig. 6.4 Sample data collection form for use in agricultural studies. (Courtesy, L.F. Curtis and A.J. Hooper.)

Table 6.1 (a) Assessments of manpower requirements for ground data collection (Courtesy, L. F. Curtis and A. J. Hooper)

Area observed (km²)	Total fields	Total observers	Date	Sampling method	Sampling density	Prior training	Progress (km h⁻¹)	Fields per hour per person
635	933	10 (working in pairs)	June/July	Line traverse (road)	One field in four along traverse	Agricultural officers familiar with area	8	22
70	341	1	May	Line traverse (road)	Continuous – all fields on traverse	Agricultural officers familiar with area	4.4	15
70	341	1	August	Line traverse (road)	Continuous – all fields on traverse	Agricultural officers familiar with area	3.8	14

Table 6.1 (b) Sample study of the time allocation in ground truth data collection

Task	Percentage of total time	
	May (45 h)	*August (46 h)*
Ancillary data collection – soil samples and farming operations	12	5
Data collection and recognition on traverses	48	52
Data checking and office compilation of data interpretation of imagery	40	43

per field could cover approximately 25 fields per day. The most successful method of estimating leaf cover in a cereal crop such as barley may be to establish a relationship between crop stage and leaf cover for a sample of fields by quadrat measurements. Data for crop stage/leaf cover relationships are shown in Fig. 6.5, leaf cover in each field having been determined by 500 point observations.

Where large quantities of ground data are to be collected and handled it is necessary to develop a system of computer storage. In these circumstances coding of data in a computer-compatible form becomes desirable. Such coding is relatively straightforward where a limited range of data is to be recorded and where class intervals are known or can be predicted. In our experience, however, it is difficult to provide unambiguous codes which will cover all, or even most, land use conditions of possible significance to imagery evaluation. There is a real risk that different ground surveyors will code the same conditions differently. Attempts to devise comprehensive coding systems can, therefore, lead to complex systems which impair speed and efficiency in field surveys. Thus, where coding is employed, experimentation is necessary to test the coding system and to train field staff.

6.4 FIELD MEASUREMENTS OF SPECTRAL REFLECTANCE

We commented in section 6.1 about the increasing need in some application areas to make very low-level remote sensing observations in order to better understand, model and monitor phenomena from higher altitude platforms. *Field spectroscopy* is one such activity.

Measurement of the spectral reflectance of different surfaces in the field environment is difficult. Yet it is preferable to make field rather than laboratory measurements since the surfaces and the conditions of illumination constructed in the laboratory cannot fully reproduce the outdoor states of reflectance as sensed by airborne or satellite systems.

Targets may be illuminated either *hemispherically*, as from a cloudy sky, or *directionally* (for one direction only), as from direct sunlight on a sunny day, although natural targets are normally illuminated by the entire hemisphere of the sky. The radiation environment is composed of two distributions of electromagnetic radiation – one incoming (*irradiance*) and the other outgoing (*radiance*). The radiation geometry of the field environment in these circumstances is shown in Fig. 6.6. In practice the positions of the source of irradiation and the sensor are both defined by two angles, the *zenith* angle (angle from the vertical) and the *azimuth* angle (measured in the horizontal plane), as shown in Fig. 6.7. In order to specify the reflectance from a target completely it would be necessary to measure

reflectance from all possible positions of the source radiation and the sensor, so that the 'bidirectional reflectance distribution function' (BRDF) could be defined. Such detailed measurements are not possible in the field environment, so an alternative to the BRDF has been sought. The alternative often adopted rests on the use of a standard reflectance panel. Since a perfect reflecting panel does not exist, a correction is made for the spectral reflectance characteristics of the panel used. When this is done a 'bidirectional reflectance factor' (BRF) can be determined which can be related to the BDRF.

The reflectance of a field target can be represented by the following functional equation:

$$f(\theta_i, \varphi_i, \theta_r, \varphi_r) = \frac{dL(\theta_r, \varphi_r)}{dE(\theta_i, \varphi_i)} \qquad (6.1)$$

where dL is the radiance per unit solid angle, dE the irradiance per unit angle, and i and r the incident and reflected rays, respectively (see also Figs 6.6 and 6.7).

An alternative to the use of standard reflectance panels is to use an upward-looking spectral sensor, which shows a dependence on the zenith or azimuth angle of the incident irradiation. Both methods can be used to obtain reflectance values in the sensor viewing elevation and azimuth. The term 'spectral indicatrix' is sometimes used to describe the reflectance characteristics as measured either over a limited range of wavelengths, or over a limited set of irradiation source positions, or over a limited set of sensor positions. However, highly specialized radiometers such as the 'parabola' instrument can scan radiance and irradiance under the control of a dedicated microprocessor to provide a very large number of measurements. Such instrumentation is likely to be used extensively in the next few decades.

Field spectroscopy is, therefore, a technique used for the measurement of spectral reflectance under field conditions. It is important for it is widely recognized that such measurements are fundamental to quantitative

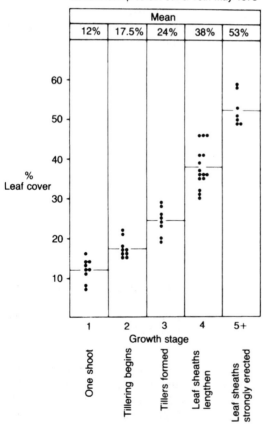

Fig. 6.5 Relationship between growth stage and leaf cover in spring barley. (Courtesy, L.F. Curtis and A.J. Hooper.)

INCIDENCE

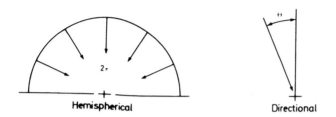

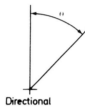

COLLECTION

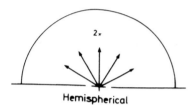

Fig. 6.6 Angular nature of reflectance measurements. A graphical description of hemispherical and directional radiation and collection. (Source: Curran, 1985.)

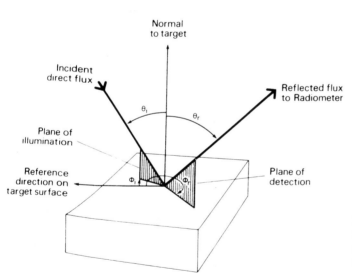

Fig. 6.7 Radiation geometry of the field environment amenable to measurement in small-site experiments. (Source: Milton, 1987.)

studies of vegetation using remote sensing. The data obtained are used in three areas of remote sensing: *calibration of data* from different platforms and sensors (or from the same sensor at different times), *prediction of best conditions* for observation and recording of data, and *modelling the reflection* from different surface structures.

The instruments used – like those now available on satellites – can be divided into those that allow the wavelength to be varied in a continuous fashion across a wide range (spectroradiometers or *spectrometers*) and those that sense a limited number of preset spectral bands (*radiometers*). However, recent advances in detector technology have blurred

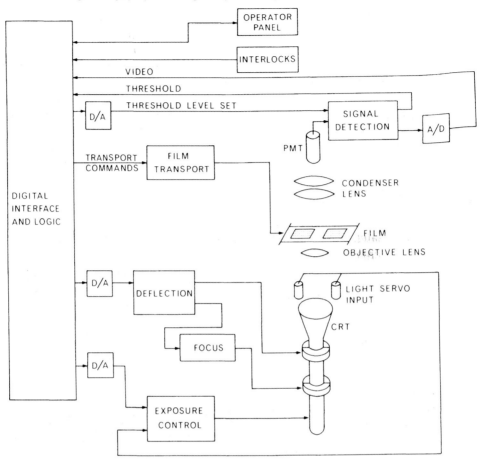

Fig. 7.4 Essentials of flying spot scanner.

example, in order to examine variations in the density of reflectance from a field containing different crops the spot size might be 50 μm. On the other hand, scanning an image for highly contrasting and coarse detail, such as rivers, may be carried out with a raster (sample frame) size of 200 μm.

Some microdensitometers use a flat-bed system in which the film is laid flat on a moving table mechanism and light is transmitted vertically through the image. However, the most common type of densitometer in use in remote sensing studies is the drum microdensitometer. Drum-type machines are usually faster than flat-bed machines and are high resolution microdensitometers where the film

is fastened onto a rotary drum and scanned by a light source when in rotary motion. A typical high-speed digital microdensitometer that has been widely used is the System P-1000 Photoscan, which provides the scanning times for a 12.5 × 12.5 cm film listed in Table 7.1.

Another method of converting film data into digital data is by scanning the film by

Table 7.1 Scan times for the P-1000 Photoscan

Drum speed (rev s^{-1})	Raster (μm)	Data rate (Hz)	Scan time (min)
2	50	14.4	20
4	100	14.4	5
8	200	14.4	1.25

means of 'flying spot' CRT (cathode ray tube) and vidicon systems (Fig. 7.4). These machines offer high speed of conversion, easy interaction with computers, possibility for data reduction by selective access to images for spectral separation, and for filtering and image improvement techniques. Flying spot scanners can read 64 levels on a grey scale and record 60 levels. They require only 20 μs for reading and 27 μs or less for recording. Spot sizes of 0.03–0.05 mm can be selected for scanning.

Traditional light sources are normally used for these scanning machines. However, exceptionally fast machines have been developed using laser beams. Laser scanners consist essentially of three components: an optical system, a film transport system and a rotating scanner system. The laser provides a high-power, collimated beam of monochromatic light which scans the film image.

Another form of converter is the electron beam recorder (EBR). The basic EBR consists of a high-resolution electron gun, an electron optical system for controlling the electron beam, a film transportation mechanism, an automatic vacuum system and regulators and electronic circuits that operate the recorder. The electron gun provides an electron spot (3–10 μm diameter) which is focused on the imagery by coils on the sides of the vacuum tube.

7.2.2 DIGITAL TO OPTICAL CONVERSION

With the development of multispectral scanning devices it is now vital to have a means whereby digital data can be converted into analog and then optical data. Such a system enables imagery (photographic or electronic) to be generated from digital data.

Generally D–A conversion is achieved by reading digital data into the computer from computer-compatible tape, transforming it into a form acceptable for digital-analog hardware equipment, and then passing the transformed data into the D–A device. Since analog

tape recorders are not able to stop and start rapidly (unlike digital recorders) it is desirable to maintain a continuous flow of data to the analog hardware. This is often achieved by setting up an input and output queue. The system is arranged so that there is always a backlog of input data waiting for processing and also a reserve (buffer) amount of data in the output to the analog hardware. This enables corrections to the input data to be made without interruption of flow to the analog device. Once the analog form of data has been generated an (optical) image can be constructed, e.g. in a visual display unit (VDU) such as a domestic television screen or computer monitor. If prints of the images are required, these can be produced either by photographing the screen or through the use of a plotting machine.

A further logical step towards the provision of digital satellite data to the user was taken by the designers of the SPOT satellite, for, as noted earlier, the pushbroom or multilinear array sensor system generates digital data directly, obviating the need for any data conversion steps to be undertaken before the observations can be processed by a computer. This approach may become the dominant one in the future.

7.3 STAGES IN THE HANDLING AND USE OF REMOTE SENSING DATA

From the above discussion the reader will realize not only that there are several different ways in which remote sensing data may be presented, manipulated and inspected, but also that there may be *different stages in the inspection and utilization of the various data types.* Indeed, these stages can be listed in the form of a sequence, leading from the original observations through to the generation of the final outputs or products required by the end-user. The following stages are those most widely recognized by the remote sensing community, although it may be noted that in some circumstances other stages may be interpolated, e.g. in the form of feedback loops or refinements.

7.3.1 DATA PREPROCESSING

This includes any treatment that has to take place before remote sensing data can be processed and analysed by the scientific user. Ideally, it should be undertaken before the environmental scientist acquires the data from a central reception facility or archive; increasingly plans are being made to ensure that in future at least some of these types of numerical treatments are carried out 'at source'. Data preprocessing commonly includes data *navigation*, *registration* and *rectification* (together often resulting in data mapped to a standard map projection), *cleaning* and *primary calibration*. Most of these procedures are carried out at central reception and/or archiving facilities.

7.3.2 DATA PROCESSING

This includes any data manipulation necessary or helpful for later scientific analysis and interpretation, but which the analyst or interpreter does not want to have to perform. Increasingly data processing is being undertaken or offered by central facilities, although in the simpler cases it is usually carried out by the user. Its chief object is to present remote sensing data in manageable forms and quantities. It often involves some element of *selection* from, and/or *compression* of, the preprocessed data, and/or *enhancement* of selected features of particular significance to the customer.

7.3.3 DATA ANALYSIS

This entails inspections of the information content of the data set in its own terms, i.e. searches for important features and repetitive patterns inherent in the remote sensing data themselves. A key element is *feature recognition*, whether based on instantaneous (single time) imagery, multitemporal or multidate imagery, or time series data, which may be of unispectral, bispectral or multispectral types. The results may or may not be immediately comparable with classifications of *in situ* data.

7.3.4 DATA INTERPRETATION

This covers the assessment of the contents of remote sensing data sets through comparison with the nature and condition of the target, wherever possible confirmed and/or calibrated by *in situ* observational information. Appropriate relationships, models and algorithms are often required for meaningful intercomparisons of these two different types of data. Results of interrelations observed in selected test areas for which both remote sensing and ground data are available (i.e. 'training sets') are often prepared as the bases for dependent *supervised classifications* of remote sensing data from conventional data-sparse areas. In the absence of analyst-specified training data, *unsupervised classifications* are used instead. Here the search is for natural groupings inherent within the remote sensing data themselves, which then may be interpretable in terms of recognizable features or classes of features in the target area(s).

Lastly it must be emphasized that more and more data preprocessing is being undertaken on the satellites themselves, a point well illustrated by the case of the onboard conversion of Landsat TM data from its raw analog form to its transmitted digital form as described earlier in this chapter – and that national and international data centres are becoming increasingly well-equipped to process data to meet a variety of customer requirements (Fig. 7.5). Thus the scientist is able to focus more attention on data analysis and interpretation, and final end-user needs, knowing that the products provided are of better quality because of improved quality control of the data on which they are based.

7.4 PREPROCESSING AND PROCESSING EXAMPLES

Before advancing in subsequent chapters to review in considerable breadth and depth the methods and results of remote sensing data analysis and interpretation activities, it is

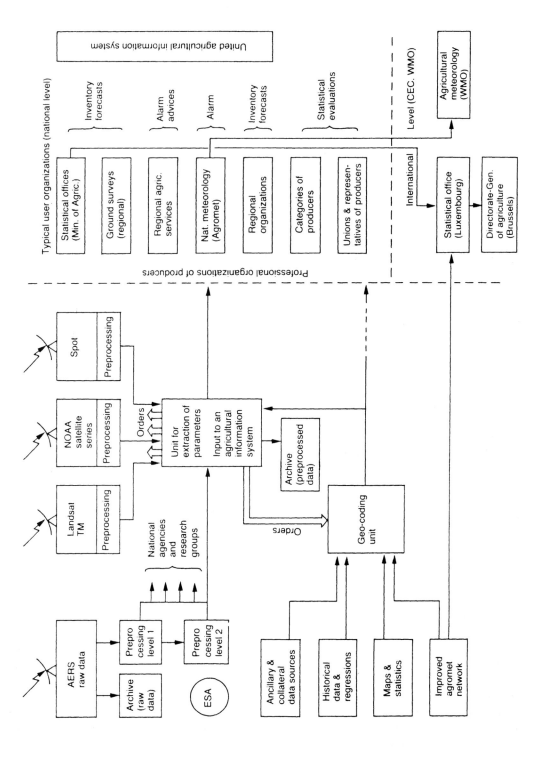

Fig. 7.5 A possible scenario for data flow from a central facility to an (agricultural) end-user community. (Courtesy, ESA.)

Table 7.2 Enhancement processing

Camera negatives	Film positives	Duplicate negatives	Intermediates	Additive printing filters	Integral colour print
N1	→ P1	→ N1′	I_1 (N1′ + P6)	→ G	→ Colour
N6	→ P6		I_2 (N7′ + P1)	→ R	→ derivative
N7	→ P7	→ N7′	I_3 (N1′ + P7)	→ B	→

The positives and negatives are printed subtractively (superimposed) in registration on to prepunched film to form intermediates. Each intermediate is made from superimposition of one positive and one negative each from a different band. Each intermediate is then printed additively with a filter so that the image of each one is transferred to only one of the colour layers in the colour film (i.e. I_1 = magenta layer; I_2 = yellow layer; I_3 = cyan layer). The resulting images, when developed from a single tri-pack film, provide the colour derivative transparency for viewing. The intermediates represent differences of reflectance between bands. Where no differences occur, the negatives and positives cancel out each other, so that no colour shows on the final enhancement. It has been found that the relative brightness and colour differences of objects appearing in the colour derivatives can be used to detect such changes as moisture content, soil density and vegetative conditions.

Photographic density slicing converts the analog, continuous-tone image into one displaying a series of steps, each representing a separate, different increment of density, completely isolated from lesser or greater densities. It is also called 'equidensitometry' or 'isodensity contouring'. AgfaGevaert makes Contour Film, a special emulsion for this purpose. Very precise and narrow density slices can be made with conventional lithographic films and developers. For example, lithographic films, such as Kodak Ortho Film 2556, can reach very high gammas when processed as recommended and can prove to be ideal for density slicing.

7.4.3 PREPROCESSING AND PROCESSING OF DIGITAL DATA FROM SATELLITES

Figure 7.8 summarizes the elements involved in the flow of data from satellites to end-users, with special reference to digital (*D*) output requirements.

As digital analysis systems have become ever more easily available and more capable, yet also affordable, so increasing efforts have been made to provide the user with remote sensing data basically in a digital form. Thus steps have been taken to ensure that key data type conversion processes can be carried out on board satellites rather than on the ground. As remarked earlier, good examples of this are given by the Landsat RBV, MSS and TM systems, which have been designed to record data in an analog form, but to transmit only digital data.

The Landsat data production facilities have had to be capable of handling very large quantities of data, as evidenced by Table 7.3, which gives some of the statistics on which Landsat ground station design criteria have been based. The structure of the wreathed image processing subsystems for the early Landsats is given in Fig. 7.9, showing the four main processing modes set up to produce photographs, video tapes, and (of greatest significance in the present context), computer compatible tapes (CCTs) containing data in digital form.

With the advent of the TM sensor system, with its even higher spatial and spectral resolutions, substantial modifications to the ground facilities were necessary. However, the process has been greatly aided by the fact that the analog-to-digital conversion is, in this case, carried out on the spacecraft itself, and

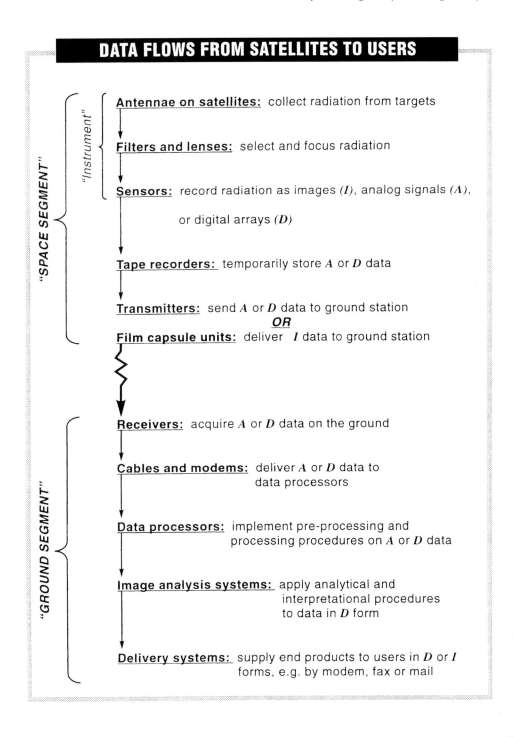

Fig. 7.8 Links between data sources and end-users via the 'Space Segment' and 'Ground Segment'.

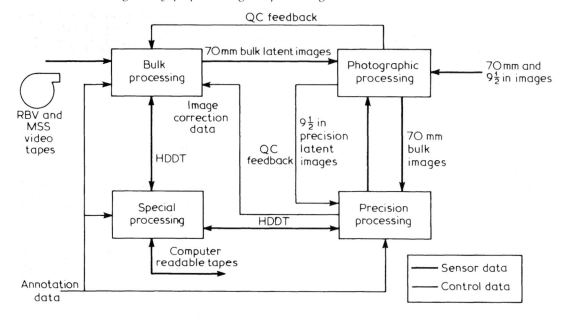

Fig. 7.9 Image processing subsystems for Landsat. (Source: NASA.)

that newer storage media (e.g. compact disks, DAT and exabyte cassette tapes) carry much more data than the now virtually defunct 1600 and 6250 bpi magnetic tapes.

Radiometrically, the TM performs its onboard analog-to-digital signal conversion over a quantization range of 256 digital numbers (8 bits). This corresponds to a fourfold increase in the grey scale range relative to the 64 digital numbers (6 bits) used by the MSS. This finer radiometric precision permits observation of smaller changes in radiometric

Table 7.3 Data processing facility requirements for Landsat Multispectral Scanner system observations (Source: NASA)

RBV input (4 bands)	Scenes per week (%)	Scenes per week	For each scene[a]	Items per week
1316 scenes/wk	100% bulk	1316	4 B-W masters	5 264
5264 images/wk			40+	52 640
			40−	52 640
			40+ prints	52 640
	20% bulk colour	263	2 C−	526
		66	20 C prints	5 260
	5% precision		4 B-W masters	264
			40+	2 640
			40−	2 640
			40+ prints	2 640
			2 C−	132
			20 C prints	1 320
			Ability to digitize	−
	5% computer readable	66	3 copies computer readable	≈713 tapes

[a] B-W, black and white; C, colour; +, positive transparency; −, negative transparency; + prints, positive paper prints; C prints, colour positive paper prints.

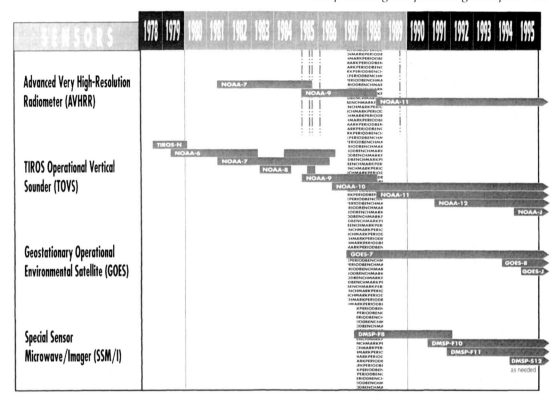

Fig. 7.10 The Pathfinder benchmark period. (Source: NOAA/NASA, 1994.)

magnitudes in a given band and provides greater sensitivity to changes in relationships between bands. Thus, differences in radiometric values that are lost in one digital number in MSS data may now be distinguished.

Geometrically, TM-4 and TM-5 data are collected using a 30 m IFOV (for all but the thermal band, which has a 120 m IFOV). Compared with the Landsat MSS this represents a decrease in the lineal dimensions of the IFOV of approximately 2.6 times, or a reduction in the area of the IFOV of approximately 7 times. At the same time, several design changes have been incorporated within the TM to improve the accuracy of the geodetic positioning of the data. Most geometrically corrected TM data are supplied using 28.5 × 28.5 m pixels registered to the Space Oblique Mercator (SOM) cartographic projection. The data may also be fitted to the Universal

Transverse Mercator (UTM) or Polar Stereographic projections.

Because of the very large amounts of data generated by a system like Landsat, it is necessary to use a computerized database to record essential data concerning each scene. The primary Landsat database, located at the EROS Data Center (EDC) in Sioux Falls, South Dakota, includes all necessary information concerning date, location and quality of all Landsat imagery. The public has access to this database through requests e-mailed or telephoned to EDC as described in the information distributed by EDC, the US Geological Survey, and other organizations (e.g. Eosat). In general, a user can also obtain, free of charge, information regarding availability of aircraft imagery, a service especially valuable to all who wish to use remote sensing imagery synergistically.

us to determine the precise nature of the local enterprise; for example, the combination of one or two tall chimneys, a large central building, conveyors, cooling towers, and solid fuel piles point to a correct identification of an installation as a coal-burning thermal power station.

8. *Resolution.* More than most other picture characteristics, spatial resolution (i.e. effectively the size of the smallest features expected to be identifiable in a particular image) depends upon aspects of the remote sensing system itself, including its nature, design and performance, as well as the ambient conditions during the sensing programme, plus subsequent processing of the acquired data. Resolution always limits the size and therefore in many cases, the nature, of features which might be recognized. Some objects will always be too small to be resolved (e.g. fair-weather cumulus clouds on the average low-orbiting weather satellite photograph), while others lack sharpness or clarity of outline (e.g. the exact position of a shoreline is often difficult to deduce from an air-photo of scale, say, 1:10 000).

9. *Stereoscopic appearance.* When the same

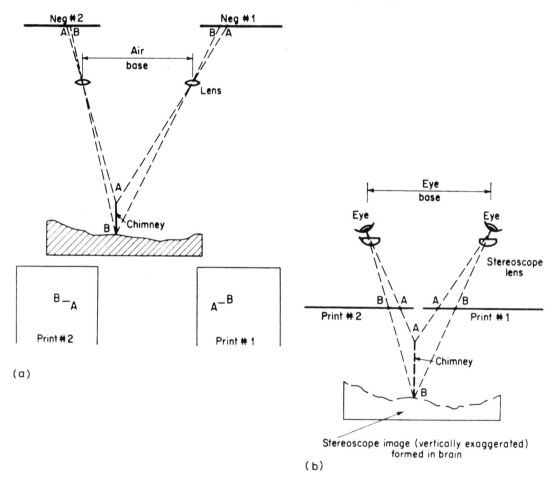

(a)

(b)

Fig. 8.2 Schematic diagrams showing: (a) the appearance of a vertical object (e.g. a chimney) on both negative and positive – note the lean of the chimney top on the prints owing to height displacement; (b) the stereo image formed in the brain by viewing the prints through a stereoscope.

feature is photographed from two different positions with overlap between successive images, an apparently solid model of the feature can be seen under a stereoscope (Fig. 8.2). Such a model is termed a 'stereo model' and the three-dimensional view it provides can aid interpretation. This valuable information cannot be obtained from a single image and is usually less easily obtained from scanner images. (Further discussion of stereoscopic images can be found on p. 157–164.)

In practice these nine elements assume a variety of ranks of importance. Consequently, the order in which they may be examined varies from one type of imagery to another, and from one type of study to another. Sometimes they can lead to assessments of conditions not directly visible in the images, in addition to the identification of features or conditions that are explicitly revealed. The process by which related, theoretically invisible, conditions are established by inference is termed *convergence of evidence*. It is useful, for example, in assessing social class and/or income group occupying a particular neighbourhood, to note features such as detached or tenement types of building forms, proportion of open space, trees, size of gardens, road patterns, traffic and proximity to industrial plants or railroad facilities (see also Chapter 17). Likewise, in estimating soil moisture conditions in agricultural areas the tonal or colour registration of vegetation, species identification, existence of field drains, riverine features, saline encrustations and land use provide contributory evidence for interpretation of the soil moisture regime (Chapter 13).

Photo-interpretation may be *regional* or *spatial* in its approach and objectives, as in the case of terrain evaluation or land classification. Figure 8.3 is a sample photo-interpretation key for land use and vegetation mapping. On the other hand, photo-interpretation may be highly *site-specific*, and related to very precise goals. In the field of transport assessments, for example, features such as car parks, railway yards or shipping berths are the subject of special study.

It is important to bear in mind that the degree of accuracy achieved in photo-interpretation may vary considerably depending on the nature of the subject, the type of photography and the skill of the interpreter. Consideration must be given, therefore, to the question of how much ground checking will be required to produce a result comparable in objective accuracy to a ground survey.

Note, finally, that *whereas a map offers classification and symbolization of the features that it records, the image does not*. This lack of symbolization makes an image more difficult to interpret but it also makes it potentially much more informative, and more flexible, for those prepared to set up their own classifications on the basis of the image characteristics.

8.3 MEASUREMENT AND PLOTTING TECHNIQUES: SELECTED ASPECTS OF PHOTOGRAMMETRY

Various aspects of plotting from scanner data and scanner images are discussed elsewhere in relation to particular types of image data (e.g. radar, infrared linescan). This section will concentrate on measurement and plotting from aerial photographs. Photogrammetry deals with the techniques involved in this specialized and well-developed science. A detailed consideration of photogrammetric techniques cannot be given in the space of this volume and so the reader is directed towards further reading (see Bibliography); only selected points need be presented here.

Aerial photographs may be either *vertical* or *oblique*, according to whether the camera axis is vertical or not. A 'high oblique' includes the visible horizon, whereas a 'low oblique' does not (Fig. 8.4). Oblique photography is most often obtained by single cameras but it is sometimes obtained by multiple camera arrangements of the 'trimetrogon' type. In this a single, vertically oriented camera is flanked

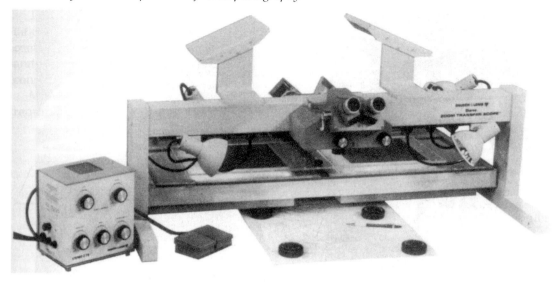

Fig. 8.10 Bausch and Lomb Transfer Scope, a graphical data transfer instrument that combines optical views of stereoscopic image and the base map. Used widely to transfer detail from photographs to maps. (Courtesy, Survey and General Instrument Co., Edenbridge, Kent.)

photogrammetric workstation exemplified by the Leica SD2000 (Fig. 8.12), in which all the photogrammetric functions (user interface, real time program, calibration, orientation, triangulation, DEM collection, etc.) are built into the control computer (a standard PC). Such a workstation is, in effect, a 'three-dimensional digitizer', capable of running any applications software for mapping, computer-aided design (CAD), or Geographic Information System packages (GIS) (Chapters 9 and 19).

When the basic procedures of photo-interpretation and photogrammetry were developing some 60 years ago, the amounts and range of image data were limited. The explosive development of remote sensing techniques generally has necessitated the evolution of a wide range of related preprocessing methods. These modern methods are designed to improve the precision of the data to be analysed, and allow surplus data to be discarded. The new generations of satellites offer far greater capabilities than aircraft could ever do, both for areas covered and for information provided, with corresponding increases in data flow. Increasingly, satellites rather than

aircraft will constitute the primary sources of data for photogrammetric analysis and map production.

To support such activities, digital workstations, in which analog or optical data are replaced by digital images, have begun to evolve since the late 1980s. These are extensions of existing digital image processing systems (Chapter 10). In them, stereoviewing of satellite imagery is made possible either by polarization techniques or by attached mirror stereoscopes. They are designed to act as universal facilities for purposes of monoplotting, stereoplotting and rectification – using digital satellite images, not air photographs, as the inputs.

One such workstation is the KERN DSP1 (Fig. 8.13). This is equipped with both image processing and analytical photogrammetric capabilities, having hardware for image movement and data recording (trackball and/or handwheels, footdisk and twin footswitches), and for stereoscopic observation (a retractable optical system to view images on a high-resolution, split-screen monitor). All common image processing functions are available

Fig. 8.12 The Leica SD2000 analytical stereoplotter. Unlike the DSP 1 and the DVP, this uses film images as input. (Courtesy, Leica.)

Fig. 8.11 The Digital Video Plotter (DVP) System. (Courtesy, Leica.)

through either hardware or software, plus an optional transputer array to enable the acceleration of time-critical functions. Photogrammetric tests of the SPOT-1 system confirmed that at least 1:100 000 scale planimetric requirements can be met thereby. Tests in France led to the generation of 1:100 000 *digital elevation models* (three-dimensional digital arrays of surface morphology) from SPOT data of the Marseilles region (Fig. 8.14) with no discernible differences between the satellite maps and existing ground survey maps. Tests of the accuracy of terrain height determinations from the SPOT products compared with those from ground survey/air photography for over 33 000 point locations yielded SPOT-derived standard errors of only ±5 m. Such results give great confidence to the ever-growing number of

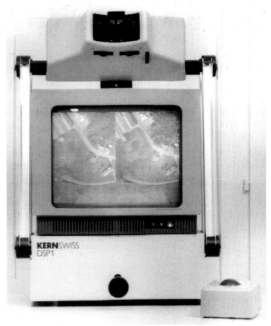

Fig. 8.13 The display and analysis system of the KERN DSP 1 Digital Stereo Photogrammetric System. (Courtesy, Leica.)

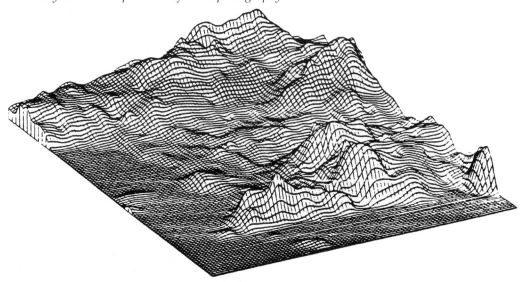

Fig. 8.14 A DEM of the Marseilles area of southern France, generalized from SPOT digital analog data.

companies and national government survey bureaux who, between them, are using SPOT data to improve – or even to generate – morphological and topographic maps of substantial areas of the world.

Thus great progress is being made with the development of analysis and interpretation systems for air photographs and Earth resources satellite imagery: clearly, SPOT in particular has provided the spur for great advancement in these fields, for digital systems offer unprecedented flexibility to the operator, not only in analysis and interpretation *per se*, but also in respect of the generation of products for use with other types of data, whether *in situ* or remotely sensed.

9.1 THE NEED FOR NUMERICAL DATA MANIPULATION

The analysis and interpretation of remote sensing data by manual means has much to recommend it. For example, the trained human eye can identify a very wide range of imaged features with both ease and accuracy. Consequently the interpretation of aerial photographs, space photographs and other forms of remote sensing imagery is, perhaps, still best carried out by hand if the data sets are small and budgets are low. But the volumes of data generated by new remote sensing systems are growing rapidly, and already far exceed the capabilities of trained interpreters in some environmental fields. For example, it has been estimated that a single low-altitude Earth-orbiting satellite, such as any of the NOAA weather satellites, has yielded something of the order of 10^{10}–10^{12} points of new data every day. Earth resources satellites now being built will be capable of even greater outputs, yielding in excess of 500 million data bits per second, i.e. between 10^{13} and 10^{14} new units of information daily. Still further increases may be expected as transmission bandwidth restrictions are eased.

Computer-based interpretation techniques afford the only possible solutions to the problem of routine extraction of useful facts from such high-volume data streams, especially on a near real-time ('operational') basis. It is therefore fortunate that, as we saw in Chapter 7, the three basic forms of remotely sensed data (image, analog and digital) are all interchangeable, for one consequence is that, if necessary, *any remote sensing data set can be rendered into a digital form and thereafter processed automatically.*

Another stimulus to the increasing use of computer analysis is the fact that numbers are required for environmental hypotheses to be tested rigorously. Although manual, 'eyeball' or *subjective* analyses can provide much useful information, and are still the better suited to some types of tasks, *objective* analyses of numerical data arrays of suitable magnitude for statistical processing can now be undertaken very quickly, and results from different areas, or different algorithms, can be intercompared. Put differently, the eye is good at *feature recognition* and *image interpretation*, but relatively poor at reliably assessing basic image characteristics such as *brightness* and *texture*. So, the choice often lies between range of features and accuracy of identification on the one hand (these benefiting from the skill and experience of the analyst), and numbers of cases, replicability and speed of operation on the other (these benefiting from computer-based automation).

To a great extent the success of 'automated' or objective techniques of image analysis is dependent upon the type of question that is posed: increasingly, as a result of growing experience, acceptable answers are now being obtained in response to the posing of reasonable questions. For example, numerical methods are especially useful for the development of probabilistic statements about points, lines or areas, and for the simultaneous use of data from more than one waveband, or basic only in data preprocessing contexts, as exemplified in Chapter 7 (Fig. 7.5), but also for dependent data processing, analysis and product generation, as shown in Figs 9.1 and 9.2. Given the already vast and often highly technical literature that has grown up in this field, our concern here is not to provide a digital techniques 'how-to' manual, as the required length of this book alone strongly forbids this, but rather to give a structural overview of the nature and use of much more detailed numerical methods in remote sensing. This will serve as a springboard both into the journal literature and, more immediately, into Part Two of this book, where applications of environmental remote sensing – many of them heavily dependent on numerical methods – are reviewed, exemplified and discussed.

It should be remembered that four steps characterize much use of remotely sensed data: *data preprocessing*, *data processing*, *data analysis* and *data interpretation* (Chapter 7). Not least because the boundaries between these different steps differ from project to project, but more because the strongest affinities are between the first two and last two pairs, we will proceed in the present context to treat digital preprocessing and processing together, followed by analysis and interpretation approaches towards the end of this chapter.

Last but not least, we will consider some increasingly important questions related to the operational implementation of remote sensing programmes, before proceeding (in Part Two) to see how remote sensing is serving the community of environmental scientists and managers today.

9.2 THE QUALITY AND QUANTITY OF REMOTE SENSING DATA

Data from remote sensing missions are rarely optimal in quality. Each remote sensing system has its own characteristic capability, particularly dependent on instrument design, altitude of the platform, and performance of the recording equipment. Other less predictable, often extraneous, influences may also adversely affect the quality of the retrieved data. Preprocessing functions are necessary to reduce or eliminate such inadequacies. Some such methods involve photographic film or deploy optical or electronic analog systems (Chapter 7). Here we must concentrate our attention upon the digital approach, which is the more dependable.

9.2.1 GEOMETRIC CORRECTIONS ('GEOREFERENCING')

Remote sensing data, whether obtained in snapshot, linescan or digital array forms, come in a variety of scales and geometries, as exemplified for aircraft imagery by Fig. 9.3, and satellites by Fig. 9.4. Often these must be rectified before they can be compared in detail with existing data. The general problem may be exemplified by studies designed to investigate the relative merits of imagery from aircraft or satellite platforms for the identification and mapping of features at a selected level of detail in a particular region. The problem is most intransigent where:

1. imaged areas are topographically rough;
2. the images are obtained from systems with broad fields of view;
3. the images are obtained from low-altitude platforms, where image displacements owing to even quite modest topographic features may be considerable;
4. systems other than framing cameras are used.

Rectification can be performed by optical and electronic means (of which the former are usually cheaper), or by digital techniques. Basically the problem involves *bringing separate data into congruence*; that is to say, ensuring that point-to-point correspondences are achieved. These are necessary, within certain limits, in many remote sensing studies if point decisions or point comparisons are intended using statistical pattern recognition techniques, e.g. the fast Fourier transform algorithm to provide

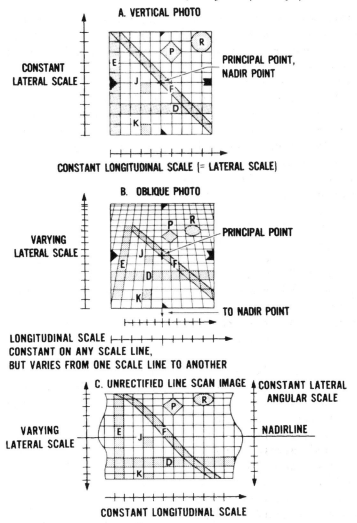

Fig. 9.3 Example of the difference in the scales and geometries of different types of remote sensing data from aircraft. The letters refer to points or areas to show effects of different geometric distortions.

the two-dimensional Fourier transform of individual images, and then to facilitate rapid cross-correlations between pairs of images or other data arrays.

Since variations in the performance of an orbiting sensor system are proportionately less than those associated with lower-altitude systems, satellite images may be brought into acceptable congruence more easily and more cheaply per unit area than those from aircraft,

using navigation and geo-registration (mapping) routines as mentioned above. The relative ease with which digital satellite data can be brought into congruence with a selected map projection is one of the principal advantages of many modern satellite remote sensing systems. Landsat is of sufficient significance today to merit special mention.

Although Landsat images are relatively free from panoramic distortions and relief displacements, they suffer from image distortions of both *systematic* (or predictable) and *random* (or unpredictable) types, for which compensations must be made before really accurate analyses can be made of these data. On the ground, each Landsat MSS or TM pixel is basically trapezoidal (not square or rectangular) in form, as a result of the general viewing angle of its sensor system, even though these effects are much more strongly marked with MSS than TM data (Fig. 9.4). Minor deviations from the norm stem from factors such as variations in the altitude, attitude and velocity of the satellite. Corrections for systematic distortions, such as the skewed-parallelogram effect of Earth rotation beneath the satellite during imaging, can be made relatively easily, in this case by offsetting each scan line by an appropriate amount.

Corrections for truly random distortions are more difficult, and are usually accomplished through reference to carefully selected *ground control points*, such as highway intersections, small water bodies, etc. For such points both image (column, row) and ground (latitude, longitude) coordinates are established, and the values then submitted to a least-squares regression analysis to determine coefficients for two 'transformation equations' which interrelate the satellite image and ground sets of coordinates. The full process by which Landsat geometric transformations are applied to the original data is termed 'resampling'. It involves:

1. the definition of a geometrically uniform 'output' matrix in terms of ground coordinates;
2. the transformation by computer of the coordinates of each output cell into corresponding coordinates in the image data set;
3. the transferral of each pixel value from the

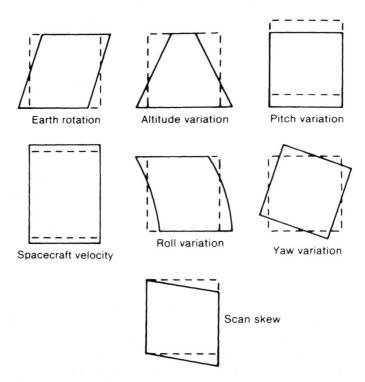

Fig. 9.4 Geometric distortions in satellite remote sensing imagery. The dashed lines indicate the desired image shape, the solid lines the actual shape. (Source: Harris, 1987.)

image data set to its appropriate place in the output matrix. The result should be a matrix of digital image data that is geometrically correct in respect of a set of ground coordinates.

9.2.2 RADIOMETRIC CORRECTIONS

These are necessary to lessen the effects of a number of variations in radiation or image density. The aim is to achieve a product with a high level of invariance with respect to the radiance of the scene, whether that radiance stems from reflected or emitted energy. Common radiometric defects are associated with the following:

1. A decline in intensity away from the centre of a lens.
2. Changing relations between the angle of view and the angle of solar illumination.
3. Variations in illumination along a sensor flight path. These are especially significant in connection with polar-orbiting satellites.
4. Variations in viewing angles of sensors scanning across the flight path of the sensor platform.
5. Variations arising from the performance of the sensor system, e.g. as they age.
6. Variations in photoprocessing and developing procedures (hard-copy imagery), and in preprocessing routines (digital data).
7. Atmospheric effects, e.g. those caused by haze, water vapour and subpixel clouds, which modify the radiation received by the sensor.

Correction of such effects is more difficult than in the case of geometric effects, for their distributions and intensities are often very inadequately known. Numerical methods of combating them include several 'normalization' techniques. Most involve the assessment of the nature and degree of the effect caused by one or more of the sources of variation in the recorded radiation density, followed by the development of appropriate physical remedies (e.g. filters or films), or numerical weighting functions (e.g. calibration curves and algorithms). In all these ways the original data may be improved – though care must be exercised, because it is not unknown for 'corrected' data to be of a lower quality than the original.

9.2.3 THE REDUCTION OF NOISE

Noise may be defined as any unwanted (either random or periodic) fluctuation of a signal which may obscure its basic form and make its analysis and interpretation more difficult. *Random noise* may be caused by the performance of remote sensing systems during recording, storing, transmission, and especially in ground reception of the data. *Periodic noise* may be caused by radio interference, and by certain components of the sensor/platform complexes themselves. In some cases it is possible to dampen random noise through the application of statistical smoothing functions to the data if these are presented in numerical form, but such noise can rarely be eliminated. Rather, efforts are necessary to keep it within acceptable bounds.

Periodic noise is the easier to eliminate because its pattern in both one-dimensional and two-dimensional products is, by definition, more systematic. When any pattern of noise is known, appropriate modifications can be made to ground station hardware to lessen its effects.

In the case of Landsat 3 an unwanted image characteristic resulted from the improper functioning of one of the six identical detectors in the Multispectral Scanner: a noticeable banding appeared with a six-line interval. Similar problems arose with Landsat TM data, but producing a 16-line banding effect because the TM scans 16 lines simultaneously and directs the energy on to different detectors. Several techniques are available to reduce the impact of these 'striping' effects (Fig. 9.5(a)). The simplest method is to multiply the pixel

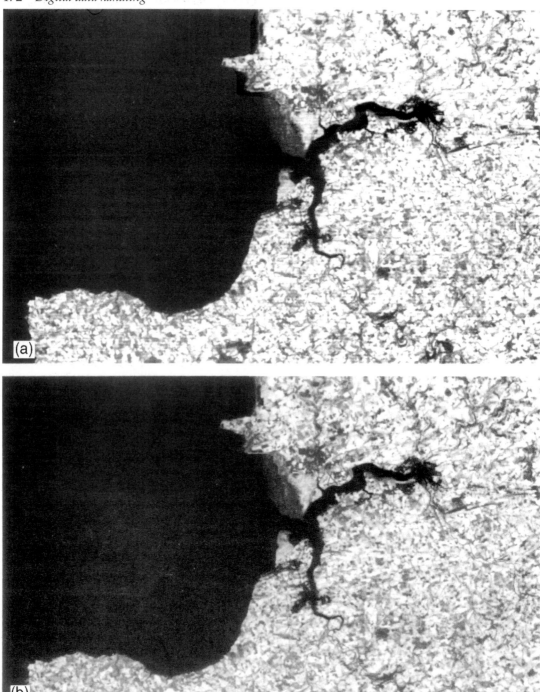

Fig. 9.5 Landsat Band 7 images of the Woolacombe area of North Devon, England, 27 May 1977. (a) Original image showing Landsat 3 striping effects. (b) Digitally destriped image. (Courtesy, Space Department, RAE, Farnborough; Crown copyright.)

values in the deficient lines by normalizing factors to give adjusted values with mean and standard deviations equal to those in the rest (Fig. 9.5(b)). Unfortunately, all such methods yield improvements that are more apparent than real: there is no way in which fully perfect adjustments can be made for fundamentally defective or missing data.

9.2.4 DATA CALIBRATION

Substantial sections of this volume are concerned with the analysis and interpretation of remote sensing data in terms of familiar characteristics of human environments. Since our funds of environmental data from conventional sources including *in situ* sensors – having been built up over a much longer period – are much greater than those of remote sensing data, it is common practice to calibrate the remote sensing data in terms of conventional parameters. However, as noted in earlier chapters, this may not always be simple or even sensible to undertake because of basic differences between remotely sensed and *in situ* data sets. However, there is now increasing interest in the analysis of the 'raw' remotely sensed radiances themselves, without attempting to combine them with some other (probably quite different) – type of data: environmental patterns mapped from satellite altitudes can be expressed simply in terms of the measured radiances (e.g. stratospheric radiances in terms of voltages), with no attempt to transform the results into more traditionally recognized environmental units. At present this approach is restricted mainly to situations in which conventional ('ground truth') data are few or unavailable, e.g. in respect of middle and upper atmospheric features (Chapter 11), but may be expected to spread in the future as remotely sensed data rather than *in situ* data become more widely accepted as the norm.

Often *prelaunch calibration* procedures are carried out to provide the numerical voltage-to-temperature or voltage-to-brightness response relations for the sensor systems. For example, the calibration routine for operational weather satellites involves the preparation of families of calibration tables for each sensor for converting normalized raw counts into effective black-body radiative temperature responses in the case of infrared sensors, or into calibrated brightness responses for the visual channel. These values may then be corrected for any bias as a function of sensor temperature, a measurement regularly available in the telemetry data.

The final step involves those corrections necessary for the interpretation of measurements of the natural Earth scene. For example, in the case of infrared sensors, allowances must be made for atmospheric attenuation, whose effects usually increase away from the sub-satellite track. This 'limb darkening' correction is generally taken as a function of local zenith angle, with the assumption that water vapour is the primary absorbing constituent. For the visual channel on current NOAA satellites the algorithm in use at present is a simple cosine function of the solar zenith angle.

9.2.5 DATA COMPRESSION

Often the full remotely sensed data set from a given platform and/or sensor exceeds the immediate needs of a particular user. The required reduction of the raw data obtained from a remote sensing system into a volume no larger than that desired for a particular programme of analysis and interpretation may be effected by one or more of a wide range of standard statistical techniques. These include *smoothing and averaging processes, sampling techniques*, and *feature extraction* methods. Smoothing or averaging processes are commonly applied by numerical means to data in either one-dimensional (e.g. line scan) or two-dimensional (data array) forms. Supplementary processing may be carried out by the computer to ensure that the assumptions of subsequent numerical processes might be met; for example, weighting factors may be applied to normalize the frequency distribution of the new, reduced population.

If the original data are to be reduced by selection instead, the principles and practices of *statistical sampling theory* are invoked. The problems of feature extraction are not difficult to resolve provided that the features to be removed for further study are related to simple characteristics of the raw data themselves – for example, regions of radiation temperatures above a given threshold in infrared studies, or areas with selected ranges of brightness in densitometric analyses of conventional photographs. Further data-reduction processes may then be applied to the information selected for careful scrutiny. However, the problems of selective feature extraction are very difficult when combinations of data characteristics are invoked simultaneously. Some of the attendant difficulties of pattern analysis and pattern recognition are discussed in section 9.3.

9.2.6 IMAGE ENHANCEMENT

Specific users of remote sensing data often require that the features of special interest to themselves be emphasized or enhanced at the expense of other (e.g. background) features. Such enhancements can be carried out photographically (Chapter 7) or digitally. The commonest digital procedures include *density slicing, edge enhancement* (image sharpening), *contrast stretching* (increasing the range of image tones), and *change detection* by image addition, subtraction or averaging. The digital approach has the advantage of flexibility, whilst the photographic approach has the advantage of cheapness (Chapter 8). The digital approaches may be further detailed as follows.

1. *Density slicing.* Given image data presented in a digital form it is possible to simplify the information by reducing the number of classes through the selection of appropriate threshold levels. Sometimes a very simple two-class classification is adequate – for example, where a categoric yes/no type of decision has to be made. Numerous digital systems have been developed for rapid density slicing. In such systems any greyscale level or levels may be selected on a photographic or electrical imprint, with output via a line printer on to a flatbed or drum plotter, or directly on to a photographic film writer. The procedure involves the development of suitable algorithms and computer programs. 'Quantization noise', or spurious contouring, may result if the slices are not chosen carefully.

2. *Colour enhancement.* Here preselected colours are accorded to chosen categories of contents in remotely sensed images, to provide more or less false colour pictures. In this way, more obvious distinctions can be made between different features of each scene.

3. *Edge enhancement.* This can be used to sharpen an image by restoring high-frequency components through the removal of scan-line noise, or to emphasize certain edges to aid interpretation by taking the first derivative in a given direction. Unfortunately the computer cannot easily distinguish desired lineaments from others of similar greyscale change, and these will be enhanced also. A common method of edge enhancement of Landsat images involves the doubling of the deviation of each pixel, or selected pixels, from the local averages. These averages can be computed from 'pixel neighbourhoods' whose shapes and sizes can be varied to suit the requirements of the data user.

4. *Contrast stretching.* Any image of low contrast or any portion of the greyscale of an image may be enhanced in digital processing through suitable adjustments to the array of picture points. Many satellite sensor systems (e.g. Landsat MSS and TM) are designed to accommodate a wide range of scene illumination conditions. Consequently the pixel values in many scenes spread over relatively narrow portions of

the full ranges that are possible. *Contrast stretch enhancements* expand the ranges of the original pixel values to increase the contrast between features with similar spectral responses, and thereby facilitate feature recognition and image interpretation. Figure 9.6 illustrates some of the available contrast stretching processes. The linear stretch (Fig. 9.6(c)) expands uniformly the observed histogram of brightness levels in a selected scene to cover the full range of available values. The *histogram stretch* (Fig. 9.6(d)) reassigns image values on the basis of their frequency of occurrence: a great number of display values (and hence more radiometric details) is assigned to the more frequently observed levels in the histogram. For more specialized analysis and interpretation, use may be made of some *special stretch* (Fig. 9.6(e)) in which features represented by a narrow range of brightness levels are greatly enhanced through the expansion of this narrow range to occupy the full range of display values, to the

exclusion of other, perhaps previously dominant, features of the scene.

5. *Special contrast stretching*. In the previous category we considered procedures designed to accentuate intensity contrasts in individual waveband data sets. Sometimes it is valuable to accentuate the contrasts between different bands of image data. One common example is the *ratio image*, which is generated by computing the ratio of digital values for two selected spectral bands for each pixel in a scene. A useful characteristic of the products of spectral contrast stretching of this type is that they are relatively free of extraneous effects, such as differential illumination (e.g. relief-induced light and shadow) across a scene. However, it is often difficult to determine which is the most appropriate choice of wavebands for a particular application. So, statistical techniques have been developed to transform pixel values into alternative sets of measurement axes so that they may be described – and later reproduced in

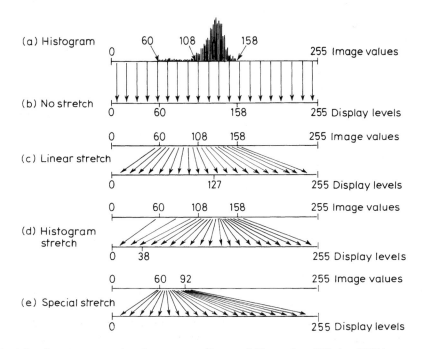

Fig. 9.6 Principle of contrast stretch enhancement. (Source: Lillesand and Kiefer, 1987.)

image form – in the most efficient manner possible for the task in hand. Such techniques are illustrated by Fig. 9.7, which relates to a simple, two-channel data set. In Fig. 9.7(a) a random sample of pixels from a selected image have been plotted according to their observed values. New axes (I and II) are introduced to provide a more efficient description of the plotted data. Such axes are characteristic of *principal component analyses*, which can be undertaken for multidimensional space: each axis or component accounts for successively smaller portions of the observed variance in the data set. In Fig. 9.7(a), axis I accounts for much more of the variance than axis II. A principal components enhancement can be generated by displaying the image in terms of the new pixel values, expressed in relation to axes I and II, not A and B as before. Where more prior information is available for a region, training set pixel values (from sites of known feature type) can be used as the basis for *canonical analysis*. In this case new axes are positioned so that the discrimination of given feature types can be maximized. In Fig. 9.7(b) three feature types can be discounted on the evidence of axis I alone.

6. *Change detection.* Change detection is often carried out by hand, using interactive man/machine methods, for example the *flicker procedure* whereby the first and second images are presented alternatively to the observer, even many times a second. Apart from simple techniques, such as the detection of greyscale change from one photograph to another, other digital procedures generally involve subtraction of an earlier image from a later one, or an addition of two such images.

9.3 THE ANALYSIS AND INTERPRETATION OF SELECTED FEATURES

9.3.1 QUANTITATIVE FEATURE EXTRACTION

Quantitative feature extraction from remote sensing data involves four steps, namely:

1. the development of methods whereby significant features or variables may be isolated;
2. the preparation of appropriate quantitative data arrays;
3. the determination of the essence of the structure of the data;
4. the reduction of the information so that only the essential point in, or structure of, the data is carried forward into the final stage of decision classification.

The first step involves the logical or *syntactical* structuring of the design of the experiment. This is the planning stage, in which considerations given to what will be measured and why; how the data will be sampled and aggregated; how many variables should be used; which methods of data retrieval, preprocessing, and data compaction will be employed; and how the final analytical stages will be organized. The second step involves some form of line-by-line data scanning to quantify the variables to be used in the analysis. These variables may be uncorrected greyscale (from a photograph, colour waveband or multispectral channel), a normalized variable, or some more sophisticated derivative from the initial data set. In most remote sensing applications the principal variables used in each channel will be tone, texture or height

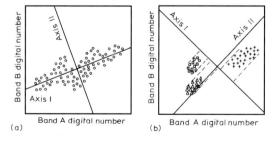

Fig. 9.7 Rotated coordinate axes used in multispectral transformations. (a) Principal components analysis. (b) Canonical analysis. (Source: Lillesand and Kiefer, 1987.)

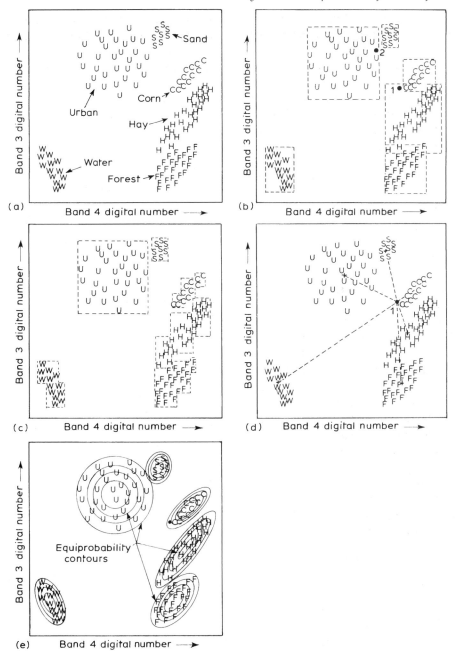

Fig. 9.8 Some common Landsat data classification strategies. (a) Pixel values from two wavebands are plotted on a bispectral scatter diagram. (b) A 'parallelepiped' classification where pixels are grouped on the basis of the highest and least class values in each band. (c) A 'parallelepiped' classification whose stepped boundaries represent each group of pixels more precisely. (d) A 'minimum distance to means' classifier where an unclassified pixel (1) is compared with established class means by computing distances between it and their centres. (e) An 'equiprobability contour' to accommodate new data, like (1), using maximum likelihood rules. (Source: Lillesand and Kiefer, 1987.)

measurements. These relate to greyscale, greyscale organization and variability, and stereoscopic characteristics. Resolution in density, or *greyscale* (the number of detectable shades of grey, e.g. 16 in a 4-bit, or 256 in an 8-bit system, etc.), may limit the subsequent stages. On other occasions limitations are imposed by the analyst – for example, considering the number of final categories required for successful completion of his or her task.

Increasingly, data from different wavebands are being combined for subsequent analysis and interpretation, e.g. by *band ratioing*, in which data from two wavebands are combined by addition, subtraction or division (Chapters 13 and 16), or in *multispectral feature space* (Chapters 12 and 17).

The third stage involves the search for key indications of the phenomena or relationships under scrutiny. Often it is sufficient to select a few aerial photographs from a large set, or the indications from some, rather than all, of the multispectral channels available. Extraneous or highly cross-correlated data can be eliminated, reducing the measurements or variables to the minimum, which is substantially orthogonal in feature space. Thus the fourth step of the extraction process is reached.

9.3.2 DATA ANALYSIS AND INTERPRETATION

In many programmes of data analysis and interpretation attention is focused either upon *point* features, or upon *non-point* features having length (lines), or length and breadth (areas). In order that image contents may be interpreted and classified correctly, it is common practice for 'training sets' of data to be compiled. These are subsamples whose identification is completely known. Where data from two or more wavebands are available it is now commonplace to consider the relations between and amongst digital values (or 'vectors') in multidimensional space. Some approaches frequently adopted in Landsat image analysis are illustrated in Fig. 9.8. Here, for the sake of simplicity, pixel values from only two bands are considered; similar processes can be applied readily to all available data bands.

In these and other ways classifications of data from known regions can be used as stereotypes of points, lines, or areas, by comparison with which unknown areas can be assessed. Thus, as we have noted before, new data may be interpreted through their resemblance to sets of templates, the known

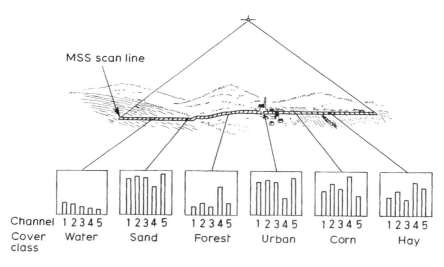

Fig. 9.9 Selected MSS measurements made along one scan line. Channels cover the following spectral bands: 1, blue; 2, green; 3, red; 4, reflected i.r.; 5, thermal i.r. (Source: Lillesand and Kiefer, 1987.)

'fingerprints' in a 'supervised' approach. The statistical parameters of each training set are calculated by analog or digital computer and can involve many variables which may have a bearing on the characteristics of the key site, ground truth station, or ground survey area. These include such factors as topography, vegetation and soil content, farming practices and environmental management as well as time of day, season of the year, and prevalent weather conditions. In calculating the statistical parameters of the training set, the greater the number of characteristics considered at a point, or the greater the number of resolution elements within an area, the better the classification accuracy may be expected to be.

Training sets are commonly used where multispectral scanner data are available. The data-display techniques that are used to facilitate feature extraction are numerous, and include histograms for each category, and by each channel (Fig. 9.9).

From such data manipulations, matrices of correlation and covariance may be compiled. The training set data usually involve *parametric discriminants*, using *Bayesian probabilities* and *maximum likelihood functions*. An extra advantage of these methods is that they commonly use thresholding procedures in which items unlike the sample are assigned to a 'discard' class, which then may be analysed in greater detail if necessary.

Many methods are being used for comparing new or unclassified data with established training sets. These include non-parametric procedures, such as *linear discriminant functions*, which are much faster in terms of computational time than the maximum likelihood decision rule, which in practice often becomes a quadratic rule requiring a large number of multiplications for each decision. Some workers have used *composite sequential algorithms* based on clustering techniques. Others have developed models based on boundary conditions. Still others have developed 'table lookup procedures', characteristically much faster than maximum likelihood functions in

practice and nearly as accurate. Operations in the spatial frequency domain have been investigated for areal analysis and classification.

Conventional training-set techniques are not without disadvantages. In particular, presupposing that detailed knowledge is not sufficiently detailed to ensure the representativeness of the training set, the ground truth sites or areas must be chosen very carefully. Furthermore, problems may be caused by variability in time and space within the training area. Consequently efforts are being made to develop techniques that are described as not supervised, for application to areas for which training sets are unavailable. Such techniques involve *clustering algorithms*, perhaps based on interactive routines whereby selected features are identified by eye and mapped. This technique relies on the fact that the multichannel radiances of different features tend to cluster in different places in the corresponding multidimensional space. The computer is then programmed to calculate the classes present, and the resolution elements that belong to them. At a later stage the classes can be identified by reference to new ground truth, or through comparisons of sample data with a bank of spectral signatures. The particular advantages of such an approach are that it is quick and cheap, no prior knowledge of the site is required, human bias is minimized, and future ground control stations can be located more objectively than in the supervised techniques.

9.4 AUTOMATIC DECISION/ CLASSIFICATION TECHNIQUES

Often in the early stages of any remote sensing programme some retracing of one's steps is unavoidable in the march towards an *operational* (i.e. routine) *feature recognition system*. In the early stages of research, most attention is accorded to image analysis rather than feature recognition. Often the best methods of data processing and analysis can be established only by trial and error, e.g. by the comparison of results obtained by different means. The expenditure of time, money and effort, and the complexity of the computer hardware and software to complete each method of approach, should be assessed in terms of the resources that would be available operationally, and in terms of the requirements of the consumer, all to be judged finally by the additional criteria of *accuracy, precision, frequency* and *resolution*.

Once the most appropriate method of data analysis has been established for any application the chief aim of most operational programmes is the correct recognition, identification, or classification of the contents of fresh imagery (Plates 2 and 3, colour section) as they become available. We have seen that numerical techniques are superior to manual or photographic techniques in that their decision and classification rules are, by definition, quantitative or statistical, whereas those for the human interpreter are judgmental and substantially qualitative. Against this there is the chief disadvantage of automatic techniques: allowances may be difficult to make for phenomena or conditions which are unevenly distributed or are present only from time to time. Such 'environmental noise' can be edited out by manual or man–machine mix (interactive) interpretation techniques, but may yield spurious patterns or results in automatic procedures.

Two particular cases in point are *background brightness*, which appears as an unwanted feature in climatological cloud imagery from weather satellites (Fig. 9.10) or, conversely, *foreground brightness* related to clouds in Earth surface investigations. In satellite meteorology cloud cover maps (nephanalyses, section 10.3.1) can be prepared by hand to show, among other things, the percentage cloud cover categorized by broad classes. These can be summed and averaged to yield mean cloud cover maps exclusive of most background brightness effects. However, it was recognized in a classic study even at an early stage in the evolution of satellite meteorology that digital techniques (Fig. 9.11) could provide acceptable results through the inclusion of a stage at which the minimum brightness level for each picture through the period in question is considered to be due to the albedo of the surface of the Earth. This level is especially high in frozen regions, over sandy deserts, and where sun glint from water surfaces is strong. By subtraction of 'brightness minima' from the daily mean brightness levels for every picture point, a value more representative of mean cloudiness itself is identified (Fig. 9.12). Unfortunately it is obviously more problematic to apply a technique like this to mapping of cloud cover on a short-term (e.g. daily) basis, or to cloud inventorying over ice, snow, or deserts. Other (e.g. multispectral) approaches are necessary for such purposes, through which information from several wavelengths is considered.

As remote sensing becomes recognized ever more widely as a useful or potentially useful tool for many types of Earth resources investigations and environmental monitoring programmes, so the number and range of related methodologies is expanding rapidly. Numerous additional examples of automatic decision/classification techniques will be introduced in Part Two of this book.

9.5 END PRODUCTS OF NUMERICAL PROCESSING

At the end of any decision-making operation involving the classification of image contents, an output is produced. This may be in one of a variety of forms, including graphical, cartographic and tabular modes:

1. *Graphical outputs*. These usually consist of histograms of the frequency of occurrence of identified categories of features, or related categories.
2. *Map outputs*. Maps are popular and common products of environmental remote

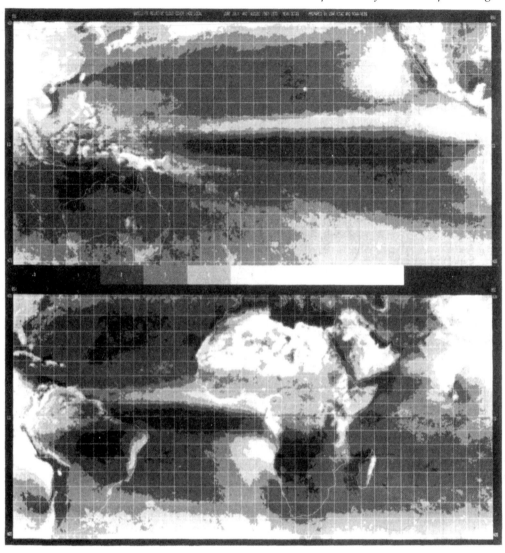

Fig. 9.10 Photographic computer maps from ESSA satellites, averaged for the season of June, July and August for a four-year period (1967–1970). The use of the Mercator projection facilitates studies of tropical brightness patterns in general and trans-equatorial links in particular. Background brightness has been retained; hence, for example, the brightly reflective desert areas of the Old World. The South Asian summer monsoon cloud over the Indian subcontinent contrasts strongly with the more typical east–west oriented ITCB clouds across the Atlantic and Pacific Oceans. (Courtesy, NOAA.)

sensing programmes, manually, interactively or automatically generated. They may be qualitative (showing types of features, e.g. crops), quantitative (showing values of features or conditions, e.g. surface brightnesses), or selected combinations of the two. The mapped patterns may be isoplethed or choroplethed; they may be single value or multiple category presentations; they may be based on absolute values, ratios, percentages, and changes through time; they may depict only point,

systems in modern atmospheric science. We shall see that by far the greatest volume of atmospheric remote sensing data obtained today is gathered from a relatively small number of weather satellites, and that their prominence in remote sensing of the atmosphere is sure to grow as larger and better equipped satellites take to orbit soon.

10.1.2 GROUND-BASED REMOTE SENSING SYSTEMS

Remote sensing of the atmosphere from the ground has been, and still is, a very fragmentary affair. This is partly because standard forecasting procedures have been developed to rely mainly on more conventional weather for their primary inputs, and partly because of the geometry of the Earth's surface.

The earliest form of artificially assisted remote sensing of the atmosphere was cloud photography from the ground. Later, balloons

and aircraft were used to provide new views of these very significant atmospheric aggregates of water droplets. The invention of films of different speeds and sensitivities, camera and radiometer filters to eliminate specific wavelengths of radiation, and time-lapse techniques in photography have all made important contributions to the meteorology of clouds. However, there are limits to the value of the essentially local information such studies of the clouds can provide. One result is that today cloud photography from the ground is dominated by the amateur enthusiast, not the professional weather observer. At principal meteorological stations three aspects of the clouds are routinely observed:

1. the proportion of the sky that is cloud covered;
2. the type or types of clouds present at different levels;
3. the height(s) of the cloud base (or bases if the cloud is multilayered).

Of these the last is the only one which is sometimes evaluated by instrumental remote sensing from the ground: the height of the cloud base may be assessed either by searchlights ('ceilingometers') or lasers.

Without doubt the most important ground-based system of atmospheric remote sensing – and that in most widespread use – is radar (Chapter 2). This active microwave system was quickly developed for weather analysis and short-term forecasting after the end of World War II. Today it is used to identify and examine a wide range of atmospheric phenomena, including raindrops, cloud droplets, ice particles, snowflakes, atmospheric nuclei, and regions of large index-of-refraction gradients. Basically, the most widely used types of weather radar consist of a *transmitter*, which produces very short bursts of power at the selected frequency; an *antenna*, which radiates the energy and intercepts each reflected signal before the next pulse of energy is transmitted; a *receiver*, which amplifies and transforms the received signals into video form; and a *display*

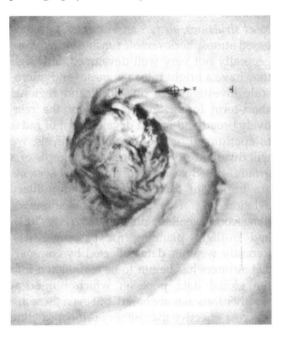

Fig. 10.1 The centre of a typhoon photographed from the lower stratosphere by a US reconnaissance aircraft. The wall clouds around the eye have a marked helical arrangement. (Courtesy, USWB.)

system, through which the returned signals can be analysed and interpreted (Chapter 4). For many purposes 'non-coherent' radar systems are adequate. These record the strength of the echo from a target, and its distribution: no account need be taken of the wavelength or frequency of the returning radar wave by comparison with the transmitted wave. But for some studies it may be important also to know the direction and rate of movement of a phenomenon with respect to the radar location. In such cases the phase of the received signal must be compared with the transmitted signal: here a 'coherent' radar system is necessary, designed to exploit the 'Doppler shift' effect. This is the observed change in frequency of radiant energy reaching a receiver when the receiver and the target are in motion relative to each other. Coherent (or Doppler) radars are now used in meteorology for a wide range of purposes. These include the measurement and prediction of turbulence and updraft velocities, e.g. at major airports; the evaluation of particle-size distributions for the local atmosphere through tracking natural scatterers; and measurements of wind speed through tracking artificial targets, such as rawinsonde balloons.

Returning to the more commonly deployed non-coherent radars, some of their principal applications are listed below, along with the more familiar types of radar display units. These include:

1. The *examination of echo intensities* in chosen directions. For such purposes the simple *A-scope* may be used (Fig. 10.2(a)). This presents the back-scattered energy in profile form, and permits the operator to compare the echo with the transmitted pulse, which also appears upon the screen.
2. The *identification and distribution of rain areas*, and the *recognition of the types of rainfall* included in them. *Plan position indicator* (PPI) display units (Fig. 10.2(b)) are in quite widespread global use for such purposes. By photographing the radar screen, or outputting its contents to a printer, or by

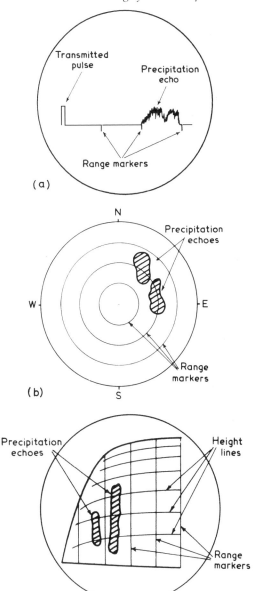

Fig. 10.2 The more commonly used forms of radar displays: (a) the A-scope; (b) the plan position indicator (PPI); (c) the range height indicator (RHI).

video-looping, at intervals, the development and movement of rain areas can be followed and predicted. PPI displays are being increasingly widely used in operational meteorology for short-term rainfall

Table 10.1 Examples of atmospheric observational programmes involving weather radar in the 1980s and early 1990s (After Collier, 1989)

Title of experiment	Operational period	Location	Precipitation measurement		Observational systems deployed	Aims
			Rain type	Measurement type		
COPT-81 Convection Profonde Tropicale 1981	May–June 1981	Northern Ivory Coast, Africa	Tropical convection	Three-dimensional structure	Mobile ground-based dual Doppler radars; surface observations; radiosondes; acoustic sounder	Development of a better understanding of the dynamic and theoretical features of precipitating convection in continental tropical regions
COST-73 European Weather Radar networking Project	1986–1991	North-western Europe including northern Sweden and Finland	All types	Mainly surface, but some three-dimensional information	Conventional and Doppler ground-based radars	Establishment of international radar data exchange
European Mesoscale Frontal Dynamics Project	Autumn–Winter 1987	South-western England and north-western France	Mid-latitude frontal systems	Three-dimensional structure	Radiosondes; dropsondes; conventional, dual polarization and Doppler ground-based radar; surface observations	Improve understanding of frontal structure and the prediction of the associated mesoscale rainfall distributions

Table 10.1 continued

The German Front Experiment	1987	Germany, Switzerland and Austria	Mid-latitude frontal systems	Three-dimensional structure	Polarimetric and conventional radar, surface and upper air observations	As for European Mesoscale Frontal Dynamics Project
GALE Genesis of Atlantic Lows Experiment	15 January–15 March 1986	East Coast, USA	Extra-tropical cyclones	Three-dimensional structure through storm life	Ground-based Doppler and conventional radars; buoys; research vessels; radiosondes; rawinsondes; surface observations including lidar; aircraft; balloon and satellite measurements	Study of mesoscale and air–sea interaction processes in winter storms with particular emphasis on their contribution to cyclogenesis
COHMEX Co-operative Huntsville Meteorological Experiment	Summer 1986	Northern Alabama and central Tennessee, USA	Subtropical convection	Three-dimensional structure throughout storm	Airborne advanced microwave moisture sounder (AMMS) (92 and 183 GHz); mobile ground-based dual-wavelength-and-polarization radar	Investigation of summertime subtropical convection
AMEX The Bureau of Meteorology Research Centre (BMRC) Australian Experiment	October 1986; January–February 1987	Northern Australia	Tropical convection–monsoon	Three-dimensional structure	Radiosondes; ground-based radar and a ship radar; surface observations	Improving understanding of the physics and dynamics of tropical weather systems

Table 10.2 A summary of important meteorological satellite families, 1959–1990

Family name	Country of origin	Number launched	Approximate period covered	Special remarks
Vanguard	USA	1[a]	Feb–Mar 1959	Early experimental satellites with primitive visible and infrared imaging systems
Explorer	USA	2[a]	Aug 1959–Aug 1961	
Television and Infrared Observation Satellite (TIROS)	USA	10	Apr 1960–July 1966	First purpose-built weather satellites
Cosmos	USSR	23[a]	Apr 1963–Oct 1983	Some weather satellites in this large, cosmopolitan Russian satellite series, including experimental operational weather satellites
Nimbus	USA	7	Aug 1964–present	Principal American R and D weather satellite
Environmental Survey Satellite (ESSA)	USA	8	Feb 1966–Jan 1972	First American operational weather satellite
Molniya	USSR	8	Apr 1966–May 1971	Dual purpose communication/weather observation satellite
Applications Technology Satellite (ATS)	USA	4[a]	Dec 1966–June 1981	First meteorological geostationary satellite to test SMS concepts
Meteor	USSR	>40	Mar 1969–present	Current Russian operational weather satellite series
Improved Tiros Observational Satellite (ITOS)	USA	1	Jan 1970–June 1971	NOAA prototype
National Oceanic and Atmospheric Administration satellite (NOAA)	USA	5	Dec 1970–Dec 1979	Second generation American operational weather satellite series
Defense Meteorological Satellite Program (DMSP)	USA	12	Feb 1973–present	Military weather satellites, often equipped with newer sensors than contemporary civilian satellites
Synchronous Meteorological Satellite (SMS)	USA	2	May 1974–Dec 1975	Testbeds for operational geostationary weather satellite systems
Global Operational Environmental Satellite (GOES)	USA	10	Oct 1975–present	First operational geostationary weather satellites
Geostationary Meteorological Satellite (GMS)	Japan	5	July 1977–present	Geostationary weather satellites covering East Asia/Pacific
Meteosat	Western Europe (ESA)	7	Nov 1977–present	Geostationary weather satellites covering Africa/Europe
TIROS-N (prototype; series numbered NOAA-6 *et seq*)	USA	12	May 1979–present	Third generation American operational polar-orbiting weather satellites
Insat	India	4	August 1983–present	Geostationary weather satellites covering India/India Ocean
Global Operational Meteorological Satellite (GOMS)	Russia	1	November 1994–present	First Russian geostationary weather satellite
Fengyun (FY)	China	2	Sept 1988–1991	First Chinese polar-orbiting weather satellites
Fengyun (FY)	China	1	June 1997–present	First Chinese geostationary weather satellite

[a] Families that have included satellites designed for non-meteorological purposes also; numbers of weather satellites only listed here.

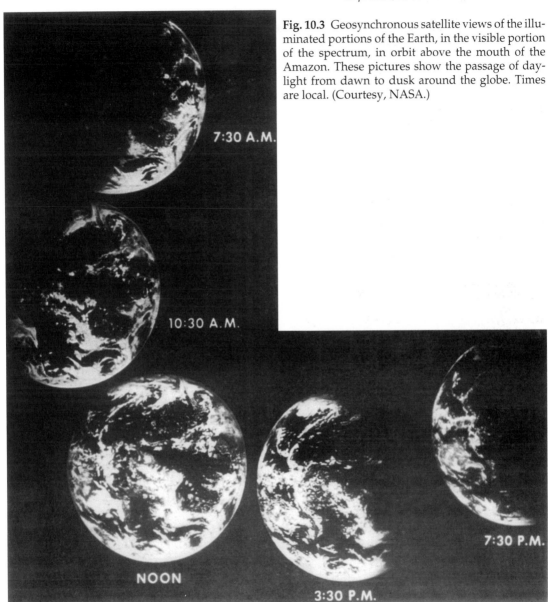

Fig. 10.3 Geosynchronous satellite views of the illuminated portions of the Earth, in the visible portion of the spectrum, in orbit above the mouth of the Amazon. These pictures show the passage of daylight from dawn to dusk around the globe. Times are local. (Courtesy, NASA.)

10.2 REPRESENTATIVE WEATHER SATELLITES

10.2.1 OPERATIONAL POLAR ORBITERS: THE NOAA PAYLOAD

Current NOAA satellites are 'third-generation' operational polar-orbiting environmental spacecraft designed to improve on and extend the functions of their predecessors, ESA 1–8 and NOAA 1–5. They are cooperative efforts of the USA, UK and France which capitalize on experience gained by these three nations in earlier satellite and sensor operations and experiments. The sun-synchronous NOAA satellites have been designed to operate in orbits inclined at 99° to the Equator, at altitudes of about 850 km, giving southbound Equator crossings between 0600 and 1000 local

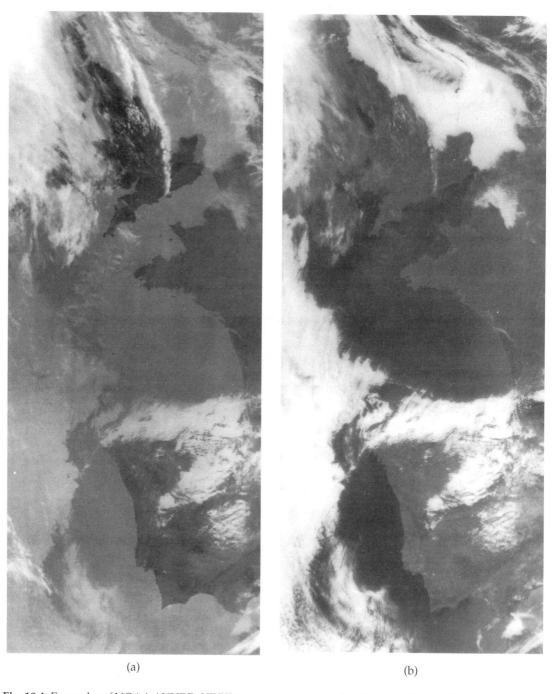

(a)   (b)

Fig. 10.4 Examples of NOAA AVHRR–HRPT images. (a) Infrared. (b) Visible, 10 May 1978. Note especially the greater sense of 'depth' in the infrared cloud field, owing to the physical relationship between cloud-top temperature and cloud-top height, which is not found in visible cloud images. (Courtesy, Department of Electrical Engineering and Electronics, University of Dundee.)

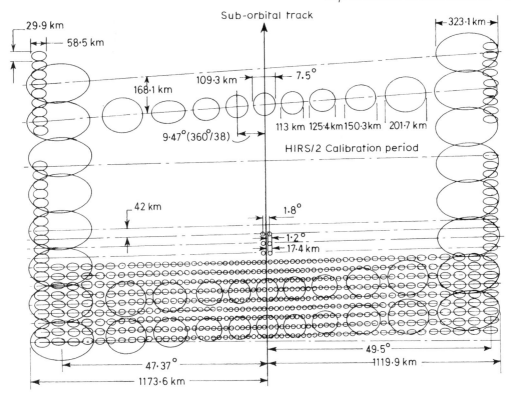

Fig. 10.5 TIROS Operational Vertical Sounder HIRS/2 (smaller element) and MSU (larger element) scan patterns projected on Earth: the microwave sensor integrates radiation over larger areas because naturally emitted target radiation is weaker in the microwave than in the infrared. (Source: Schwalb, 1978.)

solar time (LST), and northbound crossings between 1400 and 1800 LST. The most versatile instrument of the NOAA satellites is the AVHRR (Advanced Very High Resolution Radiometer), designed to provide swaths of imagery for central processing (after onboard storage) and analog-type automatic picture transmission (APT) imagery at 4 km resolution, and digital-type high resolution picture transmission (HRPT) imagery at 1.1 km resolution. The APT and HRPT imagery (Fig. 10.4) is intended for local users, and the stored data for global archiving. As mentioned earlier the AVHRR instrument is in the process of being upgraded to a six-channel instrument, and to a seven-channel ('VIRSR') instrument soon afterwards. Many countries are now able to obtain (digital) HRPT data direct from NOAA satellites; cheaper reception facilities

are required to access (analog) APT data, and such facilities are now commonplace in schools, colleges and other institutions not requiring the absolute calibration which HRPT alone provides. Meanwhile, the equally important NOAA TOVS (TIROS Operational Vertical Sounder) system comprises three complementary sensors, each providing integrated radiances from narrow columns of the atmosphere for vertical profiling purposes (Fig. 10.5). These are:

1. *The Basic Sounding Unit* (BSU), or High Resolution Infrared Sounder, designed to profile temperatures from the surface to 10 mbar, water vapour in three layers of the troposphere (both under cloud-free conditions), and the atmosphere's total ozone content.

Table 10.5 Characteristics of clouds portrayed by satellite visible images

Cloud type	Size	Shape	Shadow	Tone	Texture
Cirriform	Large sheets, or bands, hundreds of kilometres long, tens of kilometres wide	Banded, streaky or amorphous with indistinct edges	May cast linear shadows, especially on underlying cloud	Light grey to white, sometimes translucent	Uniform or fibrous
Stratiform	Variable, from small to very large (thousands of square kilometres)	Variable, may be vertical, banded amorphous, or conforms to topography	Rarely discernible except along fronts	White or grey depending on sun angle and cloud thickness	Uniform or very uniform
Strato-cumuliform	Bands up to thousands of kilometres long; bands or sheets with cells 3–15 km across	Streets, bands, or patches with well-defined margins	May show striations along the wind	Often grey over land, white over oceans, due to contrast in reflectivity	Often irregular, with open or cellular variations
Cumuliform	From lower limit of photo-resolution to cloud groups, 5–15 km across	Linear sheets, regular cells, or chaotic appearance	Towering clouds may cast shadows down sunside	Variable from broken dark grey to white depending mainly on degrees of development	Non-uniform alternating patterns of white, grey and dark grey
Cumulonimbus	Individual clouds tens of kilometres across. Patches up to hundreds of kilometres in diameter through merging of anvils	Nearly circular and well-defined, or distorted, with one clear edge and one diffuse	Usually present where clouds are well-developed	Characteristically very white	Uniform, though cirrus anvil extensions are often quite diffuse beyond main cells

that advanced forecasting systems and procedures could not do without them.

10.3.2 SPECIAL FORECASTING APPLICATIONS

In many forecasting situations some weather structures are of particular significance, because of associated extreme events. Over the years, research in many centres, but most especially in the Satellite Applications laboratory of NOAA/ NESDIS (the US National Environmental Satellite Data and Information Service), has established that a very wide range of synoptic weather systems and structures can be identified in satellite imagery, whether visible, infrared or microwave. Prominent amongst them are *vorticity patterns*: a wide range of tropical and extratropical

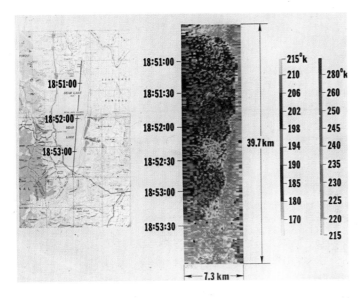

Plate 1 The flight path and microwave image (1.55 cm waveband) for 3 March 1971, Bear Lake, Utah/Idaho, from an altitude of 3400 m. (Source: Schmugge *et al.*, 1973.)

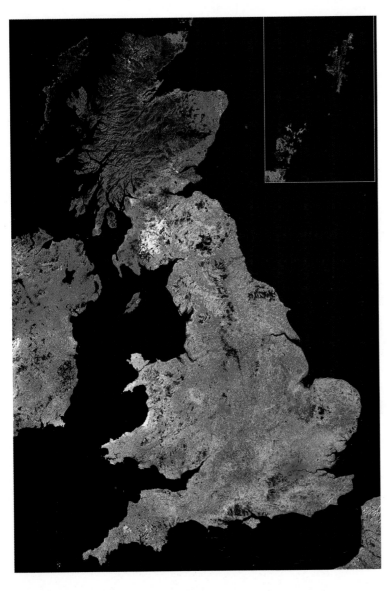

Plate 2 Composite image made by computer from Bands 4, 5 and 7 of 43 Landsat scenes. (Courtesy, BNSC.)

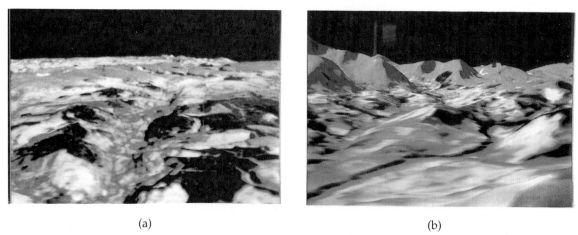

(a)   (b)

Plate 3 Example of three-dimensional image sections of the South Wales DTM (North Brecon Beacons) overlaid by co-registered Landsat TM images: (a) view to the north-east along the Vale of Neath towards the Brecon Beacons; (b) view to the south-west from Brecon to the Brecon Beacons. (Source: Barrett *et al.*, 1991.)

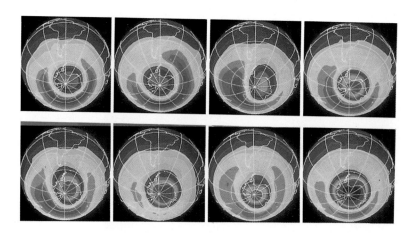

Plate 4 Development of the Antarctic 'ozone hole', annually from 1981–88, as seen by TOMS. (Courtesy, NASA.)

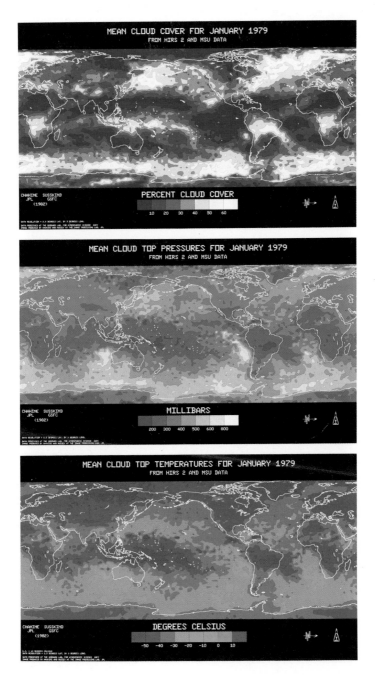

Plate 5 Global cloud climatology products derived from infra-red and microwave sounders on NOAA polar-orbiting satellites (see Section 11.3.2). (Courtesy, J. Chahine, JPL Pasadena, CA.)

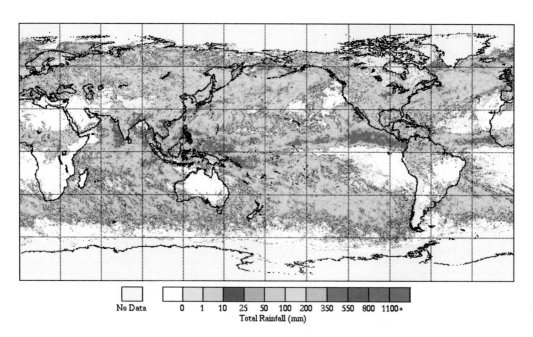

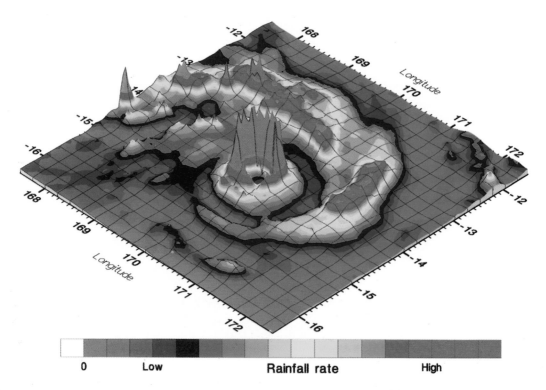

Plate 6 A global rainfall map for July 1992 based on DMSP – SSM/1 passive microwave imagery. The 'Bristol Self-calibrating' algorithm performed well in NASA's Third Precipitation Intercomparison Project (PIP-3) in 1996/97. (Courtesy, Dr C. Kidd, CRS, University of Bristol.)

Plate 7 A perspective 3D view of the field of rain rates associated with Cyclone Susan over the western South Pacific on 5 January 1998, constructed from the TMI (TRMM Microwave Imager) on the NASA/NASDA TRMM satellite. (Courtesy, Dr C. Kidd, USRA/NASA & CRS, University of Bristol.)

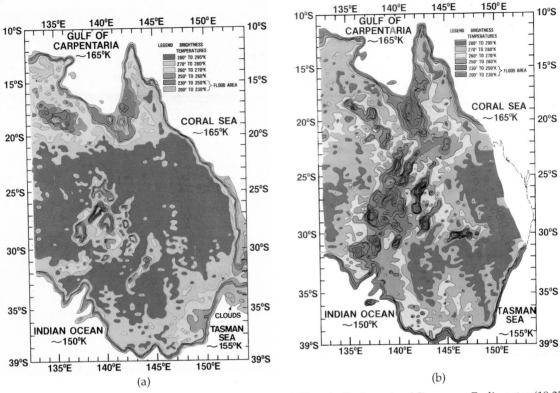

Plate 8 East Australian floods as revealed by Nimbus 5 Electrically Scanning Microwave Radiometer (19.35 GHz): (a) day, 0225-0240 GMT, 1 February 1974; (b) day, 0215-0230 GMT, 11 March 1974. Blue (250-230 K) and purple (230-200 K) surface brightness values indicate flooded surfaces. (Source: Allison and Schmugge, 1979.)

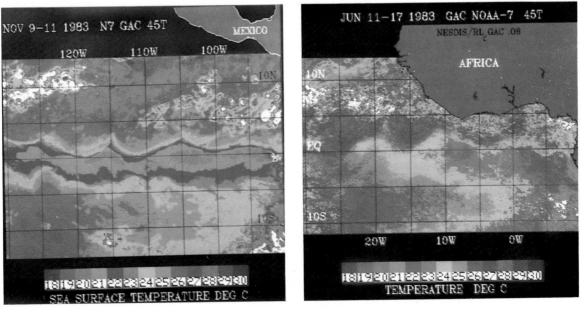

Plate 9 Colour-coded multiday maximum SST composites for the equatorial Pacific from 9 to 11 November 1983 (left), and the equatorial Atlantic from 11 to 17 June 1983 (right), from NOAA-7 AVHRR infrared data sets. These show westward moving equatorial oceanic longwaves were discovered by satellite data analyses. (Source: Legeckis, 1986.)

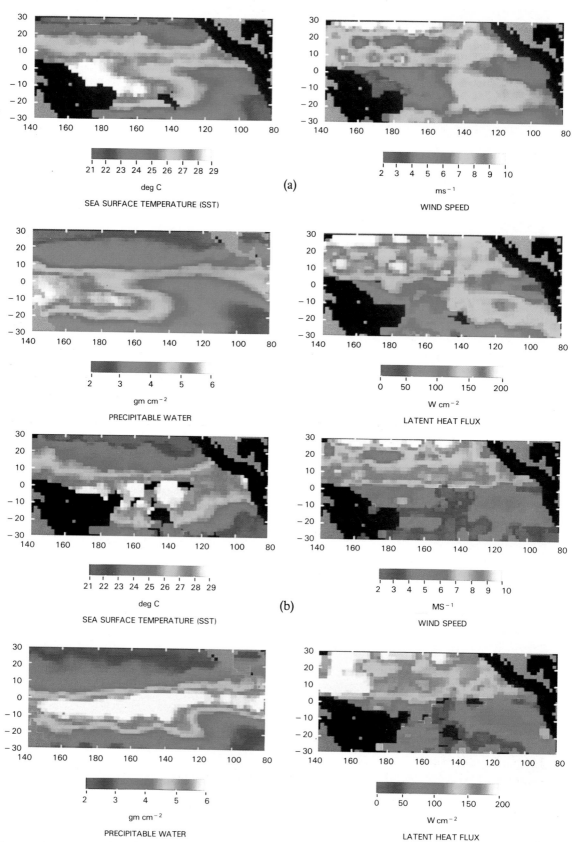

Plate 10 SMMR-derived products for (a) non-El Niño Januaries of 1980 and 1981 (averaged), and (b) the extreme El Niño January of 1983, showing four interrelated parameters. (Courtesy, NASA.)

Plate 11 Aerial infrared colour image of part of Exmoor National Park, England. Dark brown and greenish grey colours show moorland of heather, bilberry, gorse and grasses. Red tones show reclaimed fields of pasture. Pale grey fields have been cut for hay. Deciduous trees appear pink in tone. (Source: Exmoor National Park.)

Plate 12 Satellite colour image from image processor display screen showing Exmoor National Park as analysed using supervised classificatory systems. Yellow, grass moorland; purple, heather moorland; green, woodland; red, farmland. (Source: Exmoor National Park.)

Plate 13 (a) This thermal picture shows the heat loss from a large industrial complex. (b) For comparison, this shows a conventional aerial black and white photograph of the same site. The colour image depicts the variation in grey tones on the black and white print. Each colour represents a discrete temperature banding, which varies between sites, from grey and light blue (the coolest) to purple and white (the warmest).

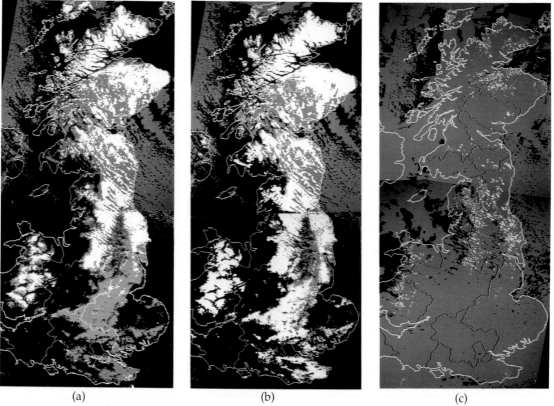

(a)   (b)   (c)

Plate 14 Snow area estimates for 27 February 1986 showing (a) partial snow cover (blue), complete snow cover (white) and cloud (grey); (b) surface temperature below 0°C (white) and above or equal to 0°C (blue); and (c) zones of melting snow (blue) and accumulating or stable snow cover (yellow). (Source: CRS, University of Bristol.)

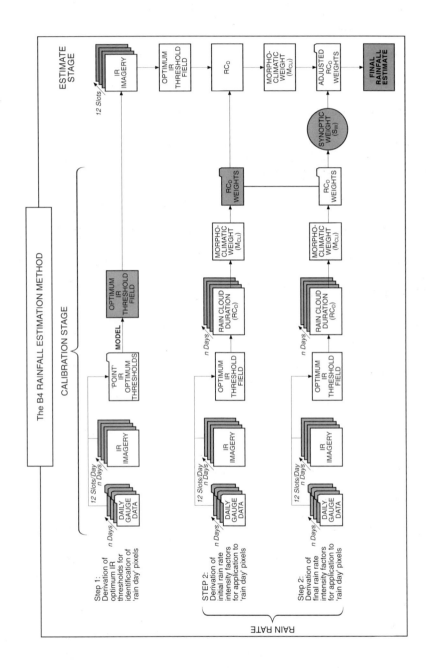

Fig. 11.8 The 'B4' (Bristol/Barrett, Beaumont, Bellerby) method for operational short-term rainfall monitoring combining satellite, raingauge, and climatological data. (Source: Barrett, 1993.)

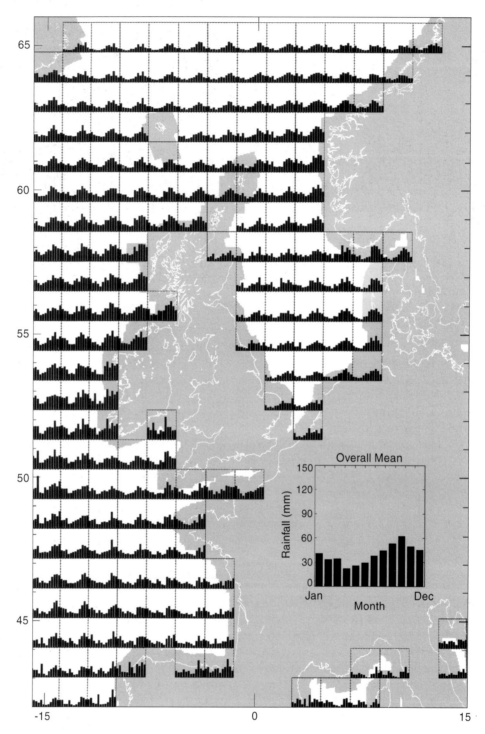

Fig. 11.9 Mean monthly rainfall over the southern and western European 'coastal oceans' derived from passive microwave satellite data calibrated by radar, from 1987 to 1996. (Source: CRS, University of Bristol.)

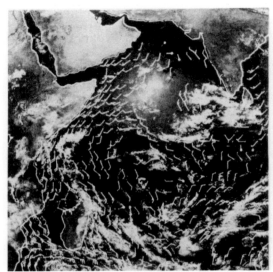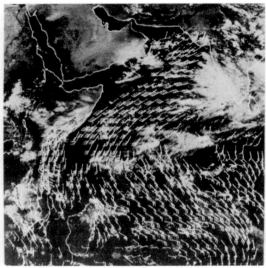

Fig. 11.10 Low-level wind fields derived from GOES-IO for (a) 17 May and (b) 21 June 1979, prepared in support of FGGE investigations. (Source: Desbois, in Tanczer *et al.*, 1981.)

It is the *cloud indexing* and *cloud climatology* types which have, in one form or another, been most widely used to date, shown most flexibility, and yielded most results in support of continuous operational rainfall monitoring programmes. For example, they have been used in support of irrigation design in Indonesia, water resource evaluation and management in Oman, desert locust control in north-west Africa, crop prediction in the Sahel, commodity forecasting in the humid tropics, and general environmental assessment in tropical Africa and the Caribbean. Experience confirms that rainfall maps based on both conventional and satellite data are more realistic in spatial detail than maps based on gauge data alone (Chapter 18), and therefore reveal more accurately the areas of average, above average and below average precipitation. Given suitable arrangements for the acquisition of input data, (e.g. from increasingly cost-effective satellite data reception facilities), satellite rainfall monitoring methods can be tailor-made to suit local needs and provide information in near real-time if required.

Today, active and passive microwave methods are becoming increasingly important, either by themselves (at least for global oceanic rainfall climatologies), or combined with geostationary infrared data, and/or radar and/or raingauge data for meteorological uses as in the WMO's 'Global Precipitation Climatology Project'. For example, the relatively physically direct but rather infrequent passive microwave data can be used to establish instantaneous rain areas and rain rates; subsequently these may be moved and/or developed using the less physically direct but much more frequent infrared data from geostationary satellites, until the next passive microwave images become available. Ideally, passive microwave imagery is needed from geostationary platforms. This, however, is not expected to become possible for many years because the weak passive microwave signals present difficult problems in sensor/platform design and engineering. More immediately, therefore, the chief aims must be to develop techniques which can be applied with equal confidence to all kinds of precipitation systems and everywhere across the globe using presently available data: despite much effort, even these have so far proved to be elusive goals, particularly in areas of surface ice and snow.

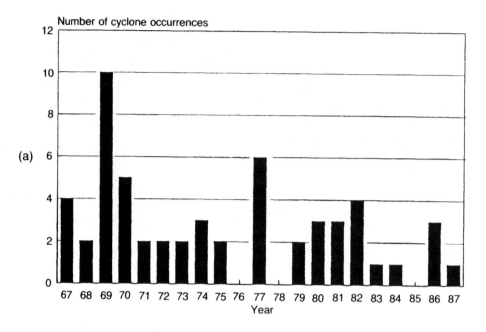

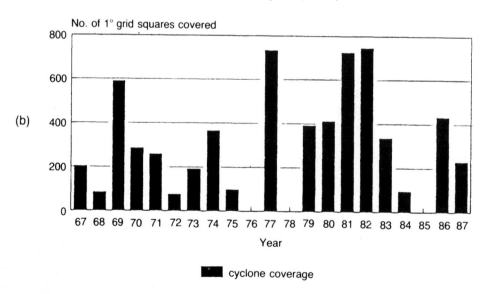

Fig. 11.11 NOAA-based climatological analyses of tropical cyclones in the Arabian Sea, 1967–1987: (a) number of storms; (b) canopy areas.

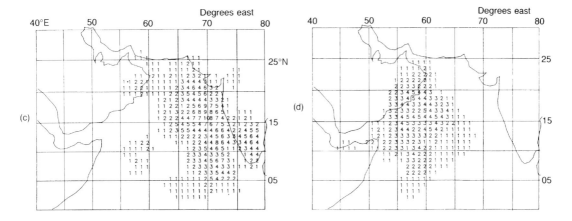

Fig. 11.11 (c) and (d) Canopy frequency of occurrence maps for two strongly contrasting years, namely 1969 (c) and 1974 (d). (Courtesy, E. C. Barrett and C. H. Power, 1990.)

11.5 WIND FLOWS AND AIR CIRCULATIONS

Several techniques have been developed for assessing the speed and direction of wind flow from satellite evidence, particularly from geostationary satellite infrared (in cloudy conditions) or water vapour channel imagery (in cloud-free conditions), although to date these have been utilized more in meteorological than climatological contexts. For example, estimates of maximum wind speeds in hurricanes are made routinely by the US Weather Bureau (Chapter 18). Cross-correlation techniques are used by major meteorological satellite data processing centres (e.g. the European Space Operations Centre at Darmstadt, Germany) to identify geostationary satellite-imaged cloud elements at selected levels in the troposphere and evaluate their direction and rate of movement from successive geostationary cloud images. Increasing interest is now being shown in these products by climatologists. Their value for air flow analyses is greatest in the tropics (Fig. 11.10), e.g. in relation to the major monsoonal systems, for such data can be processed to yield a variety of wind and circulation parameters, including wind speed and direction, relative and absolute vorticity, and horizontal divergence, especially when combined with other data (e.g. satellite cloud images themselves). Daily wind charts for tropical regions are now improved regularly through satellite inputs. Since mean monthly, seasonal and annual maps of isotachs and streamlines are prepared from such base data, satellite winds now contribute implicitly to wind and circulation climatologies, especially in some of the more conventional data-sparse regions.

11.6 THE CLIMATOLOGY OF SYNOPTIC WEATHER SYSTEMS

Remembering that it is possible to identify many atmospheric weather systems in weather satellite images, especially through their attendant cloud and radiation temperature fields, it is not surprising that many new climatological facts have emerged from studies of their time frequencies and spatial distributions through extended periods. Understandably the most interesting of such studies have been carried out for more inaccessible regions of the world, although worthwhile results have emerged even for some better known land areas, and generally in the realms of jet-stream climatology.

As mentioned in Chapter 10, particular light has been thrown upon tropical

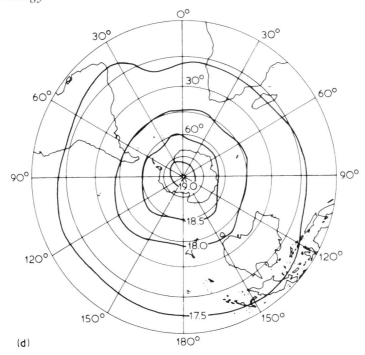

Fig. 11.13 (d) Southern Hemisphere, 4 July 1970. (Source: Barnett *et al.*, 1973.)

evidence of global climate change through the longer runs of satellite data now becoming available virtually for all levels of the atmosphere, and the world as a whole. Interest is growing in respect of possible relations amongst different climatic regions, e.g. in terms of strengths and frequencies of synoptic weather systems ('atmospheric teleconnections'), not least in relation to 'El Niño' and possible applications of such knowledge and understanding in climate prediction. These types of studies have begun to assume much greater significance with the satellite climatological confirmation and evaluation of the polar 'ozone holes', and with the rise of interest in possible global warming.

11.7 THE CLIMATOLOGY OF THE MIDDLE AND UPPER ATMOSPHERE

Many modern authorities believe that surface weather and climate are greatly influenced, or

even largely controlled, by situations and events in the upper layers of the atmosphere. Thus 'vertical interactions' are being sought with increasing vigour. Since few radiosonde balloons penetrate far into the stratosphere, pre-satellite studies of the atmosphere above the tropopause depended heavily on expensive research rocket campaigns. From these, valuable vertical profile data were obtained, but understandably their spatial coverage was very poor. Today we have a much clearer picture of the middle and upper atmosphere. Sensor systems pioneered by the CO_2-absorption waveband Selective Chopper Radiometers (SCRs, Chapter 3) of the Nimbus family provided data which have formed the bases for historic pioneering series of mean meridional profiles (Fig. 11.13(a) and (b)) and maps of selected layers high above the Earth's surface (Fig. 11.13(c) and (d)). The availability of multiyear data sets is now helping to establish the mean circulation patterns even at very high altitudes, and is revealing that the individual months exhibit a great deal of previously unsuspected variability.

Present high-altitude sounders (e.g. the MSU and SSU of NOAA satellites, as described in Chapter 10) will help to resolve some of the outstanding problems in upper atmosphere research. Prominent among these are key questions concerning the directions and degrees of vertical interactions: e.g. *does weather and climate develop principally from the top down, or the bottom up?* The answer to this may lie in the middle atmosphere (the upper stratosphere and lower mesosphere). Other key questions involve the aerosol (small particle) contents of the atmosphere, and how these may be changing – for it is known that, like clouds, these can cause substantial warming or cooling of the Earth's surface and atmosphere.

Indeed, aerosol effects in total are comparable in magnitude with those of atmospheric CO_2. Satellite estimates of both cloud and aerosol distributions in space and time will help to improve the modelling of their effects on climate variations, and in predicting future changes of climate: the first decade of the twenty-first century may well be the 'golden age' of satellite based atmospheric chemistry.

One thing is clear: through the eyes of a weather satellite system we are able to study the atmosphere much more comprehensively than before, either as a single, complex fluid continuum, or as a sum of many interrelated and more or less interdependent parts. The 1970s and 1980s were particularly exciting years for the meteorologist; perhaps the next two decades will be more so for the satellite climatologists.

12.1 THE IMPORTANCE OF WATER

Water is the most ubiquitous yet most variable of the mineral resources of the world. Alone among the constituents of our environment, water may be found simultaneously in one area as a liquid, a gas, and a solid. Unlike most other Earth resources, water (on land and in the atmosphere) is also continuously variable in its availability in one state or another. A further complication is that, geographically speaking, its patterns of concentration are always changing.

Apart from the purely scientific interest that water arouses, we have a keen practical interest in its behaviour and distribution on account of its significance as a basic requirement of living organisms, and the consequent dependence of we ourselves and our economies upon it.

For these reasons – both academic and applied – the *satisfactory monitoring of environmental water is one of the most pressing problems confronting remote sensing as a distinctive discipline today*. Aspects of it are also the most difficult.

The study of water in the environment is the preoccupation of hydrology, which has been defined by the WMO as: 'The science which deals with the occurrence and distribution of waters of the Earth, including their physical and chemical properties, and their interaction with the environment.' This modern science, along with its natural bed-fellow, water management, is growing rapidly in importance because of ever-increasing demands for adequate and suitable supplies of water for human consumption, domestic use, agriculture and livestock husbandry, even for commerce and transportation. For convenience, the areas of chief concern to hydrology and water management can be subdivided into:

1. *hydrometeorology*, which is concerned primarily with the exchange processes in hydrological cycles;
2. *surface hydrology*, which focuses on water at or near the surface of the land;
3. *hydrogeology*, which is concerned with water at or below the surface of rock and regolith;
4. *oceanography*, which deals with the nature and behaviour of the world's largest water bodies – although many would argue that this is a distinctive, even a separate, science in its own right.

We will review the use of remote sensing in each of these fields in turn.

12.2 HYDROMETEOROLOGY

The hydrological cycle of the Earth/atmosphere system can be represented in simplified form as in Fig. 12.1. This cycle is composed of three types of components: storage, transport, and exchange processes. By far the greatest storage of water is in the oceanic girdle of the globe, although significant amounts of water are retained temporarily within the rocks of the Earth's crust, in ice caps and snow fields on the surface of the Earth, in soils and regoliths, in freshwater reservoirs such as lakes and inland seas, and in the atmosphere. Water is transported by circulation systems in the atmosphere and oceans, and by movements

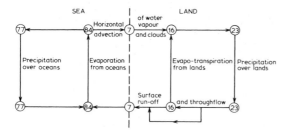

Fig. 12.1 A summary of the global hydrological cycle (assuming 100 water units).

over and near the surfaces of land areas, chiefly under the influence of gravity. It is left to the exchange processes to effect transformation of water from one physical state (solid, liquid or gas) into another. These include precipitation, evaporation and evapotranspiration.

12.2.1 PRECIPITATION MONITORING

In any review of the hydrological cycle a convenient starting point is the water supplied to land and sea surfaces by the exchange process of precipitation. Since we have already reviewed the use of satellites for improved rainfall monitoring (Chapter 11), we may focus on how surface-based remote sensing systems may be used for this purpose too, albeit in a more local and selective fashion. Microwave radar has been used for rainfall monitoring since the late 1940s. Today, operational use of rainfall radar (Fig. 12.2) is generally confined to parts of North America, Europe and the Far East, but it has a vital part to play in present and future meteorological and hydrological research, particularly as follows:

1. Radar systems can provide frequent, repetitive and real-time estimates of rainfall intensities over selected areas. For many problems, such as river management and the control of soil erosion, the short-period rainfall intensity is an important factor.
2. Radar systems can give estimates of the

total rain that has fallen over, say, a basin or a catchment through a specified period of time. Such information is often vital to water management programmes generally, and predictions of the likely flows of streams and rivers in particular.
3. Some types of radar systems can give information on the movement and development of rain cells within clouds or cloud systems ('Doppler' systems: Chapter 10). This is important in relation to our knowledge and understanding of the structure and behaviour of rain clouds and rain-cloud systems, and, therefore, to short-term rainfall forecasting.

By established practice, rainfall data are mostly obtained from networks of raingauges, preferably of a continuous-recording type. However, it is common for rainfall stations to be quite widely spaced. Especially when rain falls from convective systems (e.g. thunderstorm clouds), raingauge data may give very misleading pictures of the patterns of rainfall across wide areas. Dependent depth/area (volume) estimates of rainfall through periods of time may therefore be quite inaccurate. Shortly after the end of World War II it was recognized that radar was capable of observing the location and areal extent of rain storms, and that, in conjunction with some raingauge data for calibration, more accurate maps of rainfall might be obtainable from a suitably calibrated radar network than from raingauges alone.

If we know the technical specification and performance of a radar set, the radar equation basic to most rainfall studies may be written in simplified form as follows:

$$\overline{P}_r = \frac{cZ}{r^2} \tag{12.1}$$

where $\overline{P}_r$ is the mean received power (the 'signal') from the target, c is a constant (the speed of light), Z a reflectivity factor relating to the type of precipitation in the target area, and r the range (or distance) between the radar and

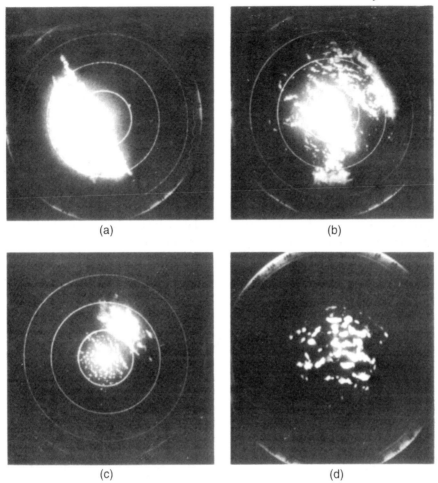

(a)　　　　　　　　　　(b)

(c)　　　　　　　　　　(d)

Fig. 12.2 Typical maritime precipitation radar patterns interpreted in terms of the cloud characteristics with which they are likely to be associated (PPI displays, 150 mile range): (a) continuous stratiform cloud; (b) ragged stratiform patches; (c) small convective cells; (d) large convective cells. (Source: Nagle and Serebreny, 1962.)

the reflecting shower of rain. Whether the output required is rainfall intensity or amount, the reflectivity factor is of particular significance: where the raindrops are large and numerous Z is high, and where they are small and few Z is low. Quantitative estimates of rainfall (R) may be derived from the target echoes by evaluating the relationship $Z = aR^b$, where a and b are local constants, empirically established. In middle latitudes, typical solutions for different types of rain are $200R^{1.71}$ for orographic rain, and $486R^{1.37}$ for thunderstorm rain. Such relations may be substituted for Z in equation (12.1) so that, for operational usage, the radar equation (in simplified form) becomes:

$$\bar{P}_r = \text{constant} \times \frac{R^b}{r^2} \qquad (12.2)$$

The implication is that for different latitudinal zones and for different types of rain *the radar signal is proportional to the rate of rainfall, divided by the square of the range.*

The principles summarized above have

Since the Chester Dee project, very significant advances have been made in implementing operational radar rainfall systems in the more developed countries of the world, including the UK (Fig. 12.5). Important advances are now being made with the development of new technology radar networks, e.g. the digital 'NEXRAD' system in the USA (Table 12.1), and the curiously named 'COST-73' system now being extended (patchily) across western Europe. However, radar costs are high and infrastructural requirements demanding; therefore, most of the land areas of the world will not be covered by weather radar in the foreseeable future. Also, because of progressive down-beam changes in the relationships between the radar

Table 12.1 Key operational goals and objectives of the US NEXRAD system (Source: NEXRAD, 1984)

1. Increase the average tornado warning time from the present 1–2 min to at least 20 min
2. Improve the accuracy of descriptions of the location and severity of thunderstorms and the ability to distinguish between severe and less-than-severe storms
3. Improve the detection of damaging winds and hail
4. Improve the safety of aircraft operation by detecting and measuring the wind shear and turbulence associated with thunderstorms
5. Provide improved rainfall estimates for flash-flood warnings
6. Reduce the size of warning areas to minimize unnecessary warnings
7. Substantially reduce the number of false hazardous weather warnings
8. Minimize the failure to detect hazardous weather due to radar outages
9. Optimize the efficacy of information provided to forecasters and other personnel by improving distribution and display of radar information
10. Detect hazardous weather conditions throughout the 50 States and at overseas locations specified by users
11. Maintain annual operations and maintenance costs at the same level as that of the radar system to be replaced (excluding the cost for radars in areas not at present covered)

signal and the height of the precipitating layer in the troposphere, the areas over which rainfall can be estimated by radar quantitatively, not qualitatively, will still remain considerably less than 100%, even in North America and Western Europe (Fig. 12.6).

12.2.2 ICE AND SNOW MONITORING

Our second important consideration in this section is frozen precipitation, especially through its surface manifestations of ice and snow. Two aspects of ice and snow distribution transcend the rest: namely, the *significance of frozen water within the hydrological cycle*, and the *influence of sea ice on maritime activities*. A substantial portion of the world's freshwater resource is stored temporarily in the form of snow, especially in high-mountain regions. Knowledge of the distribution of snow fields and their volumes in terms of 'water equivalents' is required for streamflow and water storage forecasting. In turn this strongly affects power generation and irrigation programmes, water quality control, river management, manufacturing, recreation, and many other activities which contribute to a nation's economy. Sea ice imposes a seasonal control on shipping in many regions of the Northern Hemisphere. The opening and closing of ports is influenced thereby, and certain navigation routes can be used for only a part of each year, with or without the use of icebreakers.

It is now widely recognized that satellite remote sensing has useful potential for snow mapping and monitoring (Table 12.2), and North American, Scandinavian and Alpine countries are prominent amongst those which have developed airborne techniques for estimating the volumes of snow accumulated on watersheds that serve hydro-electric power stations at lower altitudes. However, in recent years a growing number of lowland countries, including the UK, have also paid careful attention to the possibility of snow monitoring from satellites, as described below.

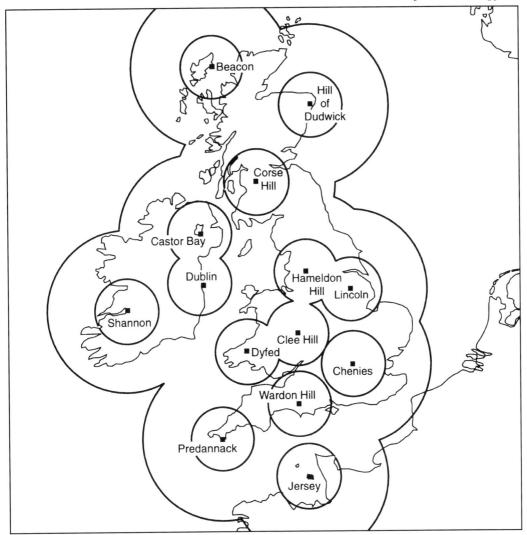

Fig. 12.6 Location of operational Meteorological Office weather radars in the UK, showing areas of quantitative rainfall estimation (inner circles) and qualitative rainfall estimation (outer circles). (Source: UK Meteorological Office.)

Several methods for snow mapping have been developed based on Landsat and local SPOT imagery to supplement or even replace surface or aircraft observations of local or regional ice and snow accumulations. Broad-scale ice and snow surveys have been based on Landsat imagery also – for example, improved maps of poorly surveyed regions of Antarctica. It has been shown that such data can be processed by computer to give snow-cover area estimates with an accuracy of 93%, or better, down to the scale of individual pixels. Highly accurate positioning of the snowline is also possible – a most important consideration in runoff predictions and monitoring of the melting process in spring and summer. Until quite recently the biggest remaining problem involved the separation of broken cloud from ice and snow. However, today's additional information from near-is quickly lost through ET. Even in humid regions, half or more of the water balance losses can be attributed to ET. Monitoring evaporation and evapotranspiration by conventional (*in situ*) means is not easy, and by remote sensing means has been described as 'The most challenging of all the applications of remote sensing to hydrometeorology' (A. Rango, 1980). The most useful satellite observations for such applications are of measurements of solar radiation, surface albedo and surface soil moisture, temperature and type of ground cover, and potentially of microwave emission and reflection too, none of which can be obtained readily over wide areas from *in situ* sources. However, observations of other parameters that affect evapotranspiration have so far proved more intransigent from satellites, including water vapour gradients and winds. Thus approaches to evapotranspiration estimation using satellite remote sensing have had to work around these missing data.

One interesting approach to evaporation monitoring from satellites has exploited the *thermal inertia properties* of soils as they are expressed through diurnal changes of surface temperatures (see also section 13.5.2). These changes are governed by the radiation budget (related to the external environment of the soil) and thermal inertia (related to the internal characteristics of the soil): the first can be modelled and evaluated with or without remote sensing inputs; the second can be assessed from infrared imagery when this is processed appropriately. We know that the thermal inertia of unsaturated soils is influenced greatly by soil porosity, and that any soil experiences a diurnal thermal inertia cycle which is closely related to its porosity. The greatest separation between the thermal inertia cycles of different soils is found about the local solar maximum and minimum, i.e. about 0200 and 1400 h local time. Much useful information relating to this problem was obtained from the Heat Capacity Mapping Mission (HCMM), whose sun-synchronous orbit was planned to image at these times. The TELL-US model was developed to yield the

following parameters from twice-daily aircraft or satellite observations:

1. thermal inertia;
2. surface relative humidity;
3. daily estimated evaporation totals, as illustrated by Fig. 12.8.

Work to achieve the simplifications which everyday operations might require is now in progress. On the other hand, exploitation of the more frequent radiation temperature measurements made by geostationary satellites such as Meteosat are likely to permit significant improvements in the accuracy of the results.

Turning next to evapotranspiration itself, within which the loss of water from vegetated as well as non-vegetated surfaces is involved, a comprehensive estimation model designed to exploit satellite data has been developed from a resistance form of the energy balance equation, in which:

$$LE = R_n - G - QC_p (T_c - T_a)/r_H \quad (12.3)$$

where LE is latent heat flux (a measure of moisture in the air), R_n is net radiation, G is soil heat flux, Q is air density, C_p is the specific heat of air, T_c is plant canopy temperature, T_a is air temperature, and r_H is thermal diffusion resistance. Key variables which can be monitored by satellites include R_n, T_c and T_a. This is very demanding, for it requires the radiation intensities observed by satellites not only to be translated into radiation temperatures, but also to be corrected for temperatures of a thin layer of the atmosphere at its very base, as with the temperature in the upper foliage of a crop or stand of natural vegetation. Since atmospheric transmission and surface emissivities are quite variable, and canopy temperatures possess a high degree of spatial variation, it is not surprising that results so obtained are thought to be too gross. It is recognized, too, that the evapotranspiration performance of a crop or vegetation area changes through the growing season. Leaf area indices (LAI, relating the leaf area of a crop to the area

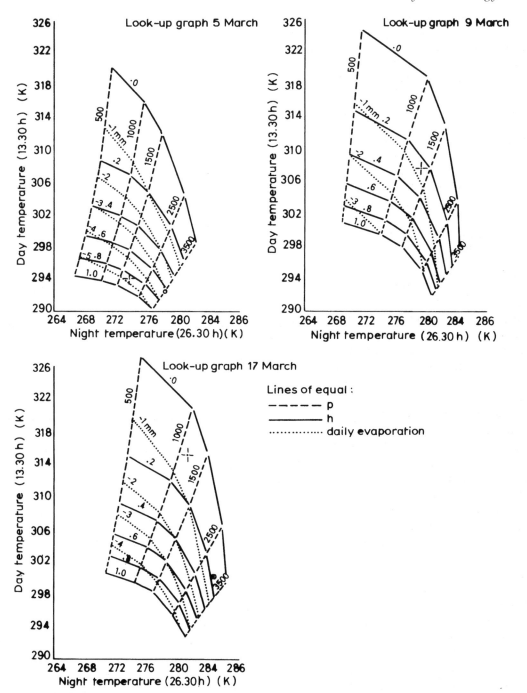

Fig. 12.8 Look-up graphs determined by the TELL-US method, representing three subsequent stages in drying of a test plot in the USA, March 1971. The crosslets mark observed day and night temperature pairs. From the graphs, values of thermal inertia, surface relative humidity and daily evaporation may be read off. (Copyright, EC.)

of the ground on which it grows) evaluated from Landsat reflectance data are being utilized in attempts to tune such methods for these seasonal effects.

The major problems that all existing remote sensing methods of evapotranspiration have yet to overcome are as follows:

1. The *process of transpiration* is still not well understood and parameterized either for structured crops, such as cereals, or for complex vegetation, such as trees.
2. In the presence of vegetation, the *surface temperature* (T_s) estimated by a thermal infrared sensor is at an unknown level within the vegetation.
3. The *most appropriate use of microwave observations* of surface soil moisture in the presence of vegetation remains to be determined.

For the future, we expect that the most practical method will probably use a multispectral approach including repetitive observations at the visible, near and thermal infrared, and microwave wavelengths. This will afford the possibility of estimating solar insolation, surface vegetative cover and/or albedo, surface temperature, and surface soil moisture from remotely sensed data and incorporating them into models of the type described here. Research continues, at present paying particular attention to the problems of how best to combine relevant point data available from the ground with areal data from the satellites, and how to develop more physically based and therefore more realistic models.

The big prize of widespread application of the successful method awaits the end of all these researches.

12.3 SURFACE HYDROLOGY

Some overlap is inevitable between this section and the subsequent chapter on soils and landforms. However, the emphases are different, for in the present context, the 'surface' is to be viewed only in respect of its significance for hydrology through run-off, infiltration and/or standing surface water. We may review this large and rapidly expanding field under six subheadings, leading us from matters of observation, through modelling, to conclude with possibilities of hydrological prediction.

12.3.1 WATERSHED PARAMETERS

The chief object of *terrain analysis* is to recognize and map 'recurring landscape features', especially through geomorphic, soil and land use characteristics. Since terrain is to a greater or lesser degree dynamic, the advent of air survey and Landsat and SPOT image analysis methods has contributed dramatically to the ability of applied hydrologists to take account of surface change factors in their assessments of significant watershed parameters. For example, deforestation and urbanization both encourage increases in the rate and volume of basin run-off, even in the absence of any change in precipitation.

In areas that are heavily clouded, recourse must be made to microwave imagery in the search for more detailed and/or more frequent information on remote and inaccessible watershed regions than could be obtained readily at ground level. A benchmark project code named RAMP (Radar Mapping of Panama) was initiated in 1965 by the US Army Engineer Topographic Laboratories to demonstrate the performance of high-resolution radar imagery in lieu of optical photography in heavily clouded regions. The sensing system was a K-band radar (see Fig. 2.3). This has a better than 99% cloud penetration capability, although it does not penetrate heavy rain or vegetative cover. Once-over coverage of the eastern end of the Republic of Panama was achieved in approximately 4 h of flying time. A number of interpretational overlaps were prepared from the final mosaic. These included surface drainage patterns, and types of surface drainage regions (Figs 12.9 and 12.10) as well as general surface configuration (topographic provinces), vegetation and engineering geology.

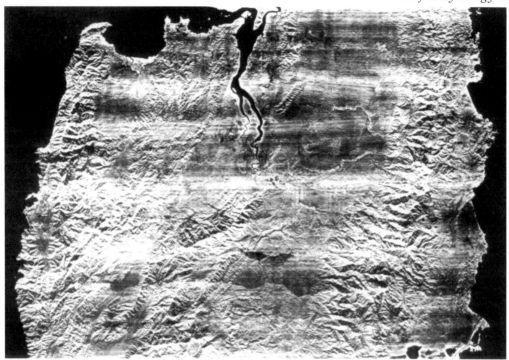

Fig. 12.9 K-band radar mosaic of Darien (Panama) and north-west Columbia. (Source: Viksne *et al.*, 1969.)

It was possible to identify variations in the local and regional expression of *surface drainage*, for example differences in the density of streams and other drainage channels, and the depth of channel incisions. In turn such variations in the drainage patterns are diagnostic of specific terrain conditions, for example the nature and depth of soil cover, lithology, morphology, and prevailing structural and tectonic influences. The patterns present themselves in an infinite variation of density and habit (or form). The *density* is determined mainly by the lithological character of the rock traversed – itself a matter of considerable interest to hydrologists – involving its hardness, porosity, solubility and consolidation. The *form* relates mainly to structure, involving faults, fractures and the attitude of bedding. Thus the drainage pattern overlay was the first to be developed from the imagery, being basic to most of the others. Similar radar mapping projects have since been undertaken in numerous regions, particularly in the tropics where

surface hydrological characteristics have been relatively little mapped, and where cloud is particularly persistent. Clearly, side-looking radar (SLR) has a part to play in regional hydrological studies, especially where cloud cover restricts the use of aircraft as platforms for conventional photography.

At the scales at which the more remote satellite systems operate most efficiently, the complexity of the total hydrological problem, and the detail required by many operational monitoring programmes, are formidable hurdles to be overcome if remote sensing is to replace more direct 'contact' methods of assessing key variables. Although the improved stereoscopic capabilities of SPOT satellites compared with Landsat have been of considerable help in such respects, this is one of the relatively few fields for which further technological development is urgently required – for example in the spatial resolution of multispectral image data, and in satellite-borne active microwave systems.

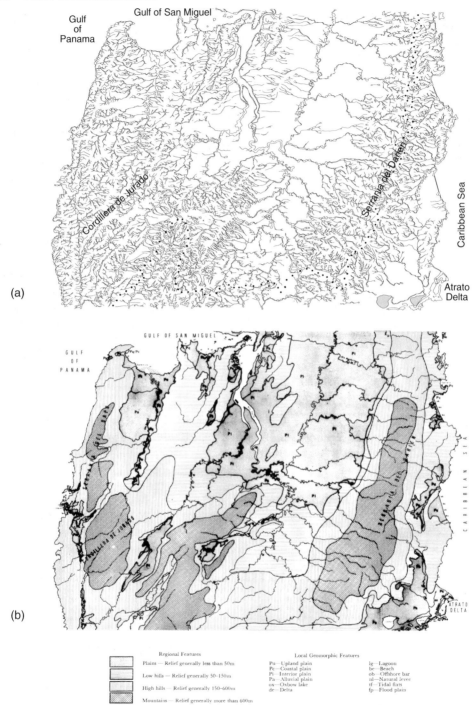

(a)

(b)

Regional Features

Plains — Relief generally less than 50m

Low hills — Relief generally 50–150m

High hills — Relief generally 150–600m

Mountains — Relief generally more than 600m

Local Geomorphic Features

Pu—Upland plain
Pc—Coastal plain
Pi—Interior plain
Pa—Alluvial plain
ox—Oxbow lake
de—Delta

lg—Lagoon
be—Beach
ob—Offshore bar
nl—Natural levee
tf—Tidal flats
fp—Flood plain

Fig. 12.10 Hydrologically significant distributions mapped from a radar mosaic of the Darien region of eastern Panama, and north-west Columbia (see Fig. 12.9): (a) surface drainage network; (b) the surface configuration. (Source: Viksne *et al.*, 1969.)

12.3.2 DRAINAGE NETWORK FEATURES

Stream network analysis, involving the mapping, measurement and classification of active drainage lines, especially in terms of stream lengths, complexities of stream patterns and the location of associated features (including springs, ponds and lakes), is an important aspect of modern hydrology. It has been shown that these can be assessed very satisfactorily from Landsat MSS imagery at scales of 1:250 000 or even better, and from TM and SPOT imagery to scales of at least 1:50 000. Furthermore, changes in dynamic features, e.g. river channel characteristics, also can be evaluated therefrom. Many rivers exhibit frequent changes of course and interactions between interlacing ('braided') channels. Even manual analyses of Landsat scenes, displayed on simple colour additive viewers, have been undertaken profitably for many regions – for example, the basin of the Kosi River in Bihar State, India, whose results revealed the following MSS band preferences:

1. Band 5: flood-plain features, e.g. abandoned channels, sandy islands, shoals, etc., and contrasts between them and nearby vegetated areas.
2. Band 7: active river courses (water is dark in the near infrared).
3. Bands 4, 5 and 7 false colour composites: alluvial areas, and classes of vegetation.

Increasingly, interactive and objective analyses of satellite imagery are now being developed for such purposes, including some based on satellite radar data, which are particularly sensitive to lineament analysis.

12.3.3 FLOOD-PLAIN MAPPING AND MONITORING OF FLOODS

Much of the world has still to be mapped on the ground at scales of 1:250 000 or better, affording much scope for Earth resources satellite imagery to be used to map important flood plain features, either for the first time or over even wider areas after each new major flood event. Unfortunately, because of their low imaging frequencies, systems such as Landsat and SPOT are not well suited to the mapping of floods, although under optimum data arrival times and sky conditions such satellites can and have been extremely valuable sources of vital flood statistics. Lesser restrictions circumscribe the use of environmental (e.g. meteorological) satellite data for such purposes, but these data carry the additional disadvantage of a relatively low spatial resolution. However, Meteosat visible and infrared imagery have been used for the monitoring of changes in large lakes and swamps (e.g. Lake Chad, and the Okavango Swamps in southwestern Africa). Passive microwave data from ESMR on Nimbus 5 were analysed successfully in studies of the widespread floods in eastern and central Australia in 1974 (Plate 8, colour section), and similar data from the DMSP-SSM/I sensors now regularly provide information on major flood events throughout the world – for example, using NOAA's 'Soil Wetness Index', based on 85H minus 19H GHz data. Data requirements for flood monitoring in basins larger than 1000 km², as proposed at a specially convened international planning meeting, are as summarized in Table 12.3. This also provides in the right-hand column an updated indication of those elements for which there is, even at present, either a significant or a reasonable satellite monitoring capability. Big improvements should become possible with the advent of EOS satellites in the next few years.

12.3.4 WETLANDS MONITORING

Here the object of attention is the hydrology of coastal and freshwater lake regions. *Surface water area* assessment is of key significance, including seasonal susceptibility to flooding and the variation of water reserves in dams and reservoirs. As remarked earlier, surface

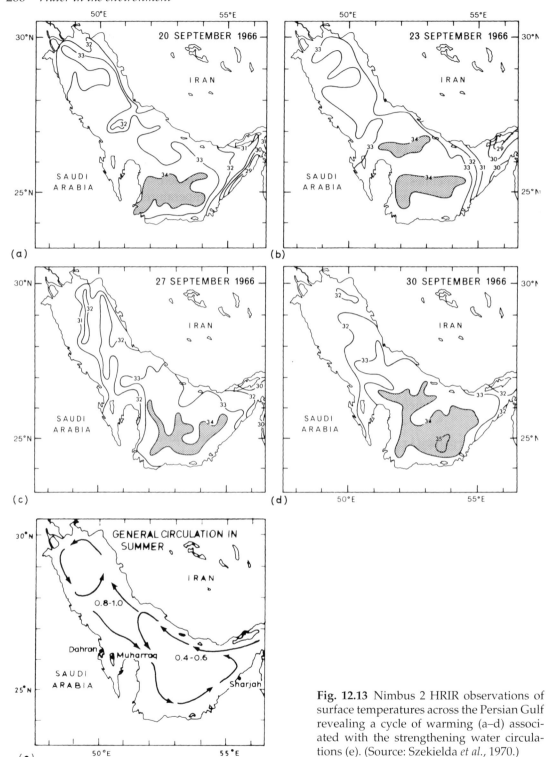

Fig. 12.13 Nimbus 2 HRIR observations of surface temperatures across the Persian Gulf revealing a cycle of warming (a–d) associated with the strengthening water circulations (e). (Source: Szekielda *et al.*, 1970.)

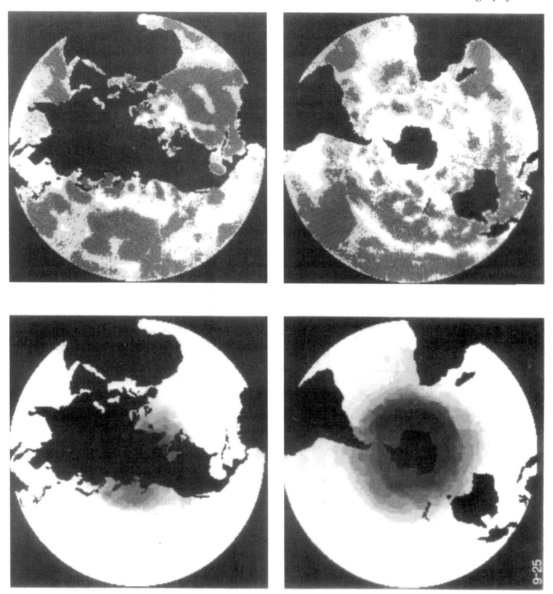

Fig. 12.14 Example of GOSSTCOMP 'quick look' photographic display of sea-surface temperatures (top) and spatial distribution of observations (bottom). (Courtesy, NOAA.)

technique applied to 1024 instrument measurements with partially overlapping fields of view in areas of roughly 100 km^2 around the retrieval point. Corrections for atmospheric attenuation are computed from VTPR data. The model develops Earth-located values of sea-surface temperature between 8000 and 10 000 times daily. Computer products include photographic displays (Fig. 12.14) and *gridded fields* (maps showing each individual temperature estimate). The accuracy of the results is checked twice daily by comparison with

temperature monitoring should be uncontaminated by the effects of clouds. Figure 12.15 illustrates some common histogram forms emerging from this work. Where cloud cover is complete, or totally absent, the histograms are characteristically *unimodal*. In partly cloudy regions they tend to be bimodal or *multimodal*. The highest temperature mode corresponds to surface temperature, the lowest to extensive cloud contamination. Analysis of remote sensing histograms is a very necessary weapon in the armoury of the remote sensing expert in this and numerous other areas of this science.

Rather like satellite cloud image analyses that have revealed features of the atmosphere which were previously unknown, satellite sea-surface temperature studies have also revealed new features of the global oceans and their circulation patterns. Examples of these are westward-moving equatorial longwave patterns observed in both the Atlantic and Pacific Oceans, which have been identified through computer analyses of AVHRR data in the 3.55–3.93 µm region, using 10.3–11.3 and 11.5–12.5 µm data from the same sensor to effect atmospheric corrections. The resulting SSTs are composited in space and time by retaining the warmest observation for each 50 km pixel in the product, so eliminating most cloud effects. The resulting wave-like features (Plate 9, colour section) have a spatial

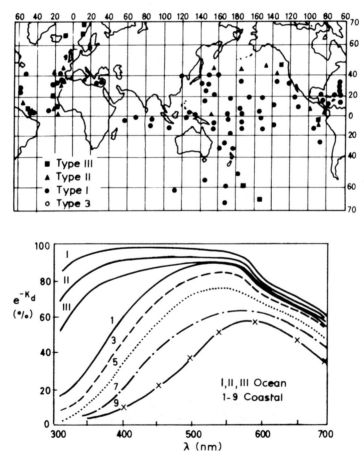

Fig. 12.16 Optical water types classified from aircraft MSS observations. (After Jerlov, 1976; source: Fraysse, 1980.)

scale of about 1000 km, and move westward with an average phase speed of 40 km day^{-1}. Since they are weak or absent in the Pacific in El Niño years, it seems likely that the waves derive their energy from the shear between the westward-flowing south equatorial current and the eastward-flowing north equatorial counter-current and the trade winds that drive them. However, differences between the waves are apparent in the Atlantic and Pacific Oceans, and more detailed studies of these differences are now proceeding.

Increasing attention is also being paid to the use of such data for *air–sea interaction* studies, which may be expected to grow in significance in the immediate future. Here El Niño itself is of prime importance, not only because of the crucial role it is thought to play in the global weather and climate system as a whole, but also because of the strong economic and human impact of recent El Niño events, e.g. the exceptionally strong El Niños of 1982–1983 and 1997–1998. This most recent event, associated with a surface temperature change of about 2°C over much of the equatorial Pacific, is reckoned to have caused thousands of deaths plus severe economic damage to South America, North America, Asia, Australia and even Africa and Europe, through associated droughts or torrential rains in different places.

Using Nimbus-7 SMMR data as the satellite information source, NASA scientists established 'normal' and 'El Niño' years, as illustrated by Plate 10 (colour section) through analyses of the January average of 1980 and 1981, and January 1983, respectively.

To examine first the 'normal' year, Plate 10(a) shows the combined monthly averages for January 1980 and 1981 of the SST, surface wind speed, precipitable water, and the latent heat flux derived from these three parameters. In this normal period, the distributions of these four parameters are closely related. The SST shows that along the Equator and away from the coasts where the SST algorithm breaks down, a cold tongue extends from the South American coast and interrupts the

warm-water belt of the tropical Pacific. Because of convective lifting of the air and surface evaporation, the regions of large atmospheric water vapour overlie the warm surface water. In contrast, the air over the cold tongue is a region of low atmospheric water vapour. Because of this surface convergence and the rising air, low wind speeds occur in these same regions. Finally, the latent heat flux is approximately proportional to the wind speed.

Later in 1982, one of the most intense El Niño/Southern Oscillation episodes began. Plate 10(b), which gives the corresponding distributions for January 1983, shows the changes that have taken place by the height of El Niño. The SST shows that the cold tongue along the Equator has vanished, owing to the disappearance of coastal upwelling, and that the entire Pacific is warmer. Examination of the winds shows a strengthening of the trade winds north of the Equator and a weakening of the winds to its south. Atmospheric convection above the warm water yields a region of large water-vapour content that extends all along the Equator. Surface observations showed that this change in the water vapour distribution led to much higher than average rainfall and flooding in certain of the equatorial islands. Finally, the latent heat flux pattern is strongly altered, with the largest fluxes now lying entirely north of the Equator. The change in these fluxes is dominated by the change in the surface wind patterns.

12.5.2 WATER QUALITY AND SALINITY ASSESSMENTS

Turning next to phenomena that are perhaps more significant and certainly more amenable to study at much smaller scales, the *turbidity* or 'murkiness' of water is affected by a wide range of natural and anthropogenic factors, including suspended sediment content, biological activity (e.g. the abundance of plankton), pollution, etc. Entire subdisciplines have grown up to study water quality (see Table 12.4), especially through reflectance properties of, and light propagation in, water bodies: *optical limnology* deals with freshwater lakes, and *optical oceanography* with oceans and seas. The theory of water optics has been exploited variously in the remote sensing of water quality. Of fundamental importance is the general rule that the spectral distribution of upwelling light from a water body contains quantitative information on the content of suspended and dissolved materials. We may consider two of the many possibilities that depend upon this principle.

First, multicategory classifications of oceanic and coastal water types have been developed from aircraft multispectral imagery, as exemplified by Fig. 12.16. This case reveals a trend of water clarity extending from the murky inshore waters of high to middle latitudes to the clearest areas of the Earth's tropical oceans. Coastal waters generally are much more turbid than open oceans, not only because of higher concentrations of pollutants but also because of large quantities of so-called *Gelbstoffe* or 'yellow substance', which is the overall product of dissolved organic matter brought by rivers into these shallow water areas. Since the design of Landsat was optimized for land applications, not for use over oceans and seas, MSS and TM data are of limited use for the mapping and monitoring of water quality types, though MSS Band 5 has proved useful for broad-scale optical wave type determination.

Second, the capacity for useful oceanic application of the Coastal Zone Color Scanner (CZCS) on Nimbus 7 may be illustrated through reference to monitoring of chlorophyll (phytoplankton) concentrations, as exemplified by Table 12.5. Pre-operational tests suggesting that CZCS visible data should be suitable for estimating chlorophyll through known relationships between chlorophyll concentrations and CZCS band differences, e.g. the difference between albedos observed in Band 1 minus Band 2 have been amply confirmed by analyses of the CZCS data themselves. Work on this topic is now proceeding

Table 12.5 Wavebands investigated by the Nimbus 7 CZCS

Band number	Wavelength (mm)	Application
1	443	Multispectral detection of chlorophyll, suspended sediment and Gelbstoffe
2	520	
3	550	
4	670	
5	750	Comparison with Landsat MSS Band 6
6	11.5 (μm)	Ocean surface temperatures mapping

with CZCS data, spurred by the knowledge that fishery advisories stand to benefit greatly from successful results, and by the expectation that new sensors of a similar type are now operational, e.g. on board the SeaWiFS spacecraft (section 5.12.2).

Finally, a word on *marine pollution monitoring*, a theme of growing international significance at the present time. Contamination of water bodies by human activities often lends itself readily to investigation by remote sensing techniques since the effluents of commercial and industrial activity often contrast sharply with the waters which receive them. Aircraft have been used in many instances to identify and map discharge patterns into the sea from sewage outfalls and industrial complexes. Oil spillages, from both ships at sea and oil leaks from maritime wells, can be located with particular ease from aircraft (including helicopters, both light and major coastal patrol types) – and potentially from satellites. There are four reasons why this is so:

1. The emissivity for petroleum products is significantly higher than for a calm sea surface.
2. Crude oil pollutants have increasing with emissivities increasing gravity.
3. The radiometric response of oil varies little with time of day or the age of the pollutant.
4. It is possible to design reliable microwave systems to map oil pollution.

Already, remote sensing data – particularly

from airborne sensors – have been used in courts of law to obtain convictions of those responsible for significant pollution events. Here and elsewhere, use of remote sensing in support of international law is becoming commonplace.

12.5.3 SATELLITE OCEANOGRAPHY: THE WAY AHEAD

The first purpose-built oceanographic satellite, Seasat, was launched into a near-polar circular orbit at an altitude of 800 km on 26 June 1978. Unfortunately the satellite ceased to provide data only 106 days later on 9 October, thus marking the end of its useful life. However, Seasat had already demonstrated the great value of its instrument package, and had returned a rich supply of both land and sea-surface data to Earth. Some researchers have remarked that the early demise of Seasat was fortunate in that it has forced the principal investigators to examine a manageable quantity of data thoroughly – rather than an overwhelming flood of data very partially, which has been the rule, not the exception, with most other satellite data sets.

Seasat carried four microwave sensors and an auxiliary Visible–Infrared Radiometer (VIRR). The suite of microwave sensors, which gave the spacecraft its all-weather operating capability, included:

1. a *radar altimeter* looking vertically down to measure wave height and the microtopography of the ocean surface;
2. a *synthetic aperture radar* (SAR), which scanned a 100 km wide swath located some 250–350 km to the right of the subsatellite track, primarily to measure wave length and direction, to provide data for the study of coastal processes, and to chart ice fields and leads;
3. a *scatterometer* scanning out to 1000 km on either side of the spacecraft for wind speed and direction assessments;
4. a *passive scanning multichannel microwave*

radiometer (SMMR), which recorded naturally emitted microwave radiation from the target at five frequencies between 6.6 and 37 GHz, primarily for global measurement of sea-surface temperature.

It was generally agreed that the additional primary objectives of the Seasat mission were well met. Many bonuses also accrued, for example through the realization that much of the SAR imagery over land contained information of potential value for geological, hydrological, glaciological and even (a useful bonus) some land use studies. Figure 12.17(a) illustrates the high resolution achieved by the Seasat SAR, and the wealth of detail in the imagery over land and in coastal zones. One exciting and entirely unexpected finding was that, under suitable conditions of sea-bed topography and tidal flow, seafloor features have visible expressions on the water surface. Figure 12.17(b) is a good example of this effect in the English Channel. Many other interesting phenomena can also be seen in Seasat SAR data, e.g. wave refraction and diffraction in coastal zones, current features, inlet plumes, and wetland characteristics.

Since Seasat, other new satellite/sensor systems have been developed which are of special significance to oceanography. Admittedly MOS-1, the first of Japan's Earth observation programme satellites, was designed with both land and marine applications in mind: its objectives were to establish a basic technology common to both areas of application, as well as to data collection systems. However, the chief emphasis of this family has been on observation of the sea surface through visible, infrared and microwave radiometers. The instrument package on MOS-1 consisted of:

1. a *Multispectral Electronic Self-scanning Radiometer* (MESSR), a four-channel visible and near infrared radiometer providing a swath width of 200 km and a resolution of 50 km (cf. the Landsat MSS);
2. a *Visible and Thermal Infrared Radiometer* (VTIR), with one visible and three thermal

Fig. 12.17 (a) Seasat SAR image of the Tay Estuary, Scotland, 19 August 1978, showing many coastal and land surface features.

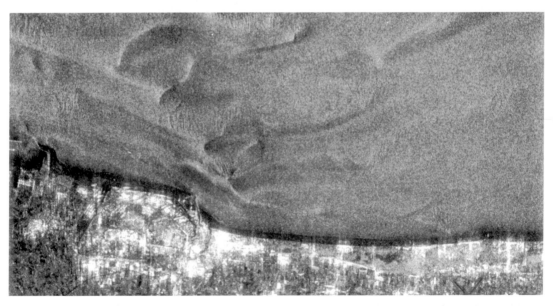

Fig. 12.17 (b) Seasat SAR image of Dunkirk, France, 19 August 1978, showing sea surface features related to the topography of the sea bed. (Courtesy, Space Department, RAE, Farnborough; Crown copyright reserved.)

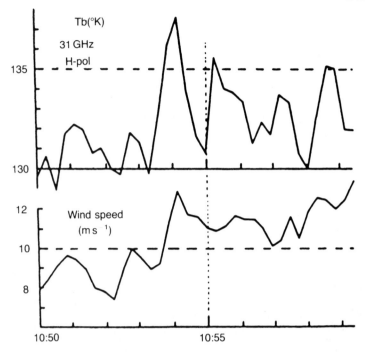

Fig. 12.18 Brightness temperature (Tb) versus wind speed plots from the MSR on MOS-1. Both are 10 min averages for data recorded on 12 November 1982. (Source: Tsenchiya *et al.*, 1987.)

infrared (one water vapour and two split window) channels, a swath width of 500 km and a resolution of 0.9 km (cf. the NOAA–AVHRR);

3. a *Microwave Scanning Radiometer* (MSR), a dual-frequency (23.8 and 31.4 GHz) vertically and horizontally polarized imaging system giving a swath width of 317 km and resolutions of 32 and 23 km respectively.

The primary applications of these instruments in oceanography include coastal zone mapping and monitoring, sea-surface temperature monitoring and sea-state evaluation (wind speeds near the surface), respectively. Pre-launch simulations of MSR evaluations of sea state are shown in Fig. 12.18. These were based on the expectation that ocean surface roughness and both microwave reflectivity and emissivity are related to one another. One consequence is that, at some frequencies and polarizations, useful relationships are found between microwave brightness temperatures

and wind speeds near the surface; some emissions from calm surfaces are strongly polarized, whilst at the same frequencies emissions from less calm surfaces are relatively unpolarized.

These principles have been more thoroughly exploited in relation to data from the wide-angle DMSP–SSM/I sensor, and instantaneous wind speed maps are prepared routinely from such data both for local regions and for the world's oceans as a whole. Claimed accuracies are of the order of $\pm 2\,\mathrm{m\,s^{-1}}$, although further algorithms benefit from local regional calibration. Wind direction products can also be generated from SSM/I data, though for weeks or months rather than instantaneously. Clear opportunities exist for such products to be used profitably in a wide range of contexts, from global air–sea interaction modelling at one extreme to fisheries and ship routeing at the other. The same sensor is also proving very useful for sea ice assessment and monitoring, and related practical applications in high-latitude and fringe polar regions.

Last but not least, reference must be made again to ERS, the first of which, after long delays, was launched successfully on 16 July 1991. Much of value for oceanography is being derived from the ERS series radar altimeters and wind scatterometers, and SAR data are also important for mapping and monitoring, especially in coastal zones and in areas of sea ice. Towards the end of 1997 the Canadian Radarsat was used to accomplish the first complete active microwave map of Antarctica, and additional data were collected for stereoscopic and interferometric analyses of glaciers and glacial ice motion, clearly revealing patterns of ice deformation and flow (Fig. 12.19). A follow-up survey has been planned for 1999, so that the two can be used to chart changes in the ice sheets and glacier fields with a comprehensiveness and precision not possible before. Meanwhile, SSM/I data already provide some

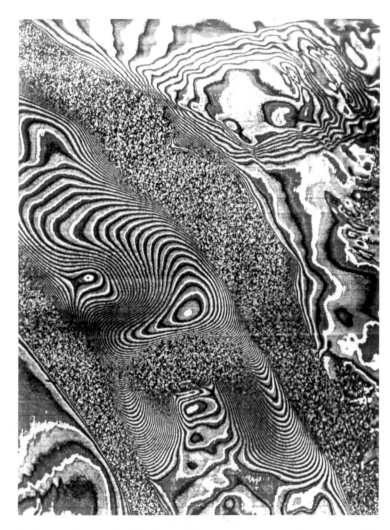

Fig. 12.19 ERS-1 SAR 'interferogram' of the Rutford Ice Stream, Antarctica generated from a pair of radar images separated in time, and superimposed on one another to reveal patterns of ice movement, and rates of flow (e.g. tighter packing of the contours on fringes of the diagonal ice stream. (Courtesy, JPL, Pasadena.)

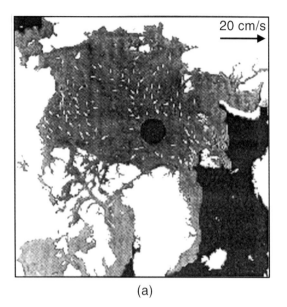

20 cm/s

(a)

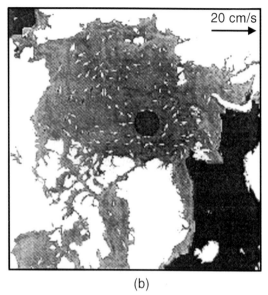

20 cm/s

(b)

Fig. 12.20 Arctic Ocean ice displacement vectors for (a) 12 December 1992 and (b) 12 January 1993. In this SSM/I image, white arrows indicate displacements derived from feature tracking using wavelet analysis; dark arrows indicate displacements from Arctic Ocean buoy positions.

broad-scale evidence of ice drift (Fig. 12.20), the most comprehensive yet available from any source.

As in other Earth-observation application areas, perhaps the most effective oceanographic uses of visible and infrared data, including SeaWiFS, and data from active and passive microwave systems, such as the ERS-1 SAR and the MOS and DMSP passive microwave radiometers, is now arising from their combined uses, the relatively high spatial but low temporal resolution data from the active systems being used for local studies and for calibrating the relatively low spatial but high temporal resolution data from the passive systems, which in turn are of most value for global applications.

Undoubtedly oceanography will benefit greatly from future dedicated satellites and instrument packages. Indeed, it is not too much to expect that oceanography – potentially, if not presently, the biggest subdivision of hydrology – will undergo a revolutionary growth in scale and practical significance as ocean-monitoring satellites increase our knowledge of maritime and maritime-margin conditions by a veritable 'quantum leap'. Since oceans cover seven-tenths of the global surface, yet have received relatively little attention compared with land areas, no one could suggest that they do not deserve much more attention in remote sensing than they have yet been given.

13.4 REMOTE SENSING OF SOILS AND LANDFORMS BY PHOTOGRAPHIC SYSTEMS

13.4.1 CONVENTIONAL AIR PHOTOGRAPHY

We have seen that engineers and agriculturists adopt different definitions and classifications of soils. As a result, remote sensing techniques are required to serve several purposes based on different definitions of soil and different basic concepts. The first major post-war study of the recognition of soil conditions, made at Purdue University, USA, was aimed mainly at identification of soil parent materials and was chiefly of value to those interested in soils engineering. It was based on two assumptions: first, that once photographic characteristics for soils at one site had been determined, the identification of similar sites elsewhere on the photographs could be used to identify similar soils; and second, that by study of aerial photographs such features as landforms, surface colours (or tones), erosion, slope, vegetation, land use, micro-relief and drainage patterns could be used to deduce the general character of the soil.

Table 13.2 Terrain unit classification

Unit	Description
Land element	The simplest part of the landscape; uniform in lithology, form, soil and vegetation
Land facet	Consists of one or more land elements grouped for practical purposes
Land system	Recurrent pattern of genetically linked land facets
Land region	Consists of one or more occurrences of land system local forms which are generally contiguous
Land division	An assemblage of surface forms expressive of a major continental structure (morphotectonic)
Land zone	The world extent of a major climatic type

Further development of this latter approach by the International Training Centre for Aerial Survey in Holland led to a widely used technique for pedological analysis. In this, each element of the landscape is mapped out separately (e.g. slope, vegetation, land use) and maps are prepared from a composite of the element boundaries. The boundaries that were coincident with more than one element of the landscape were given additional weight in the final interpretation of soil boundaries.

The widespread recognition of the close association of soils and landforms promoted much interest in *landscape mapping* as a means of reconnaissance soil mapping. Research was based on the premise that if detailed characteristics for one terrain unit were known, one could apply them to other analogous terrain units elsewhere. The problem facing terrain analysis was (and continues to be) one of classification and definition of the units to be recognized in aerial photographs. It was necessary to have some means of identifying each terrain unit and to store data relevant to the unit so defined, so that other workers could recognize them elsewhere. Thus a system of hierarchical classification was proposed, as shown in Table 13.2.

This 'land system' approach has been widely applied, for example, in Australia by Land Research and Regional Survey teams, in South Africa by engineers, and in Africa by agriculturalists. It is best applied to those areas that display distinctive and well-defined facets of landscape. Where units are ill-defined and/or where cultivation has largely destroyed the natural vegetation cover it is less easily applied.

The development of computing techniques has subsequently led to such investigations of the parametric description of landforms to be placed on an entirely numerical footing. Computer analysis of landform now seems to be within our grasp and future developments of the land system approach are likely to be in the form of automated cartography using some form of supervised classification based on knowledge of soil–landscape relationships.

Meanwhile, the accuracy of soil mapping from the air photograph has proved to be very variable. Comparisons of soil maps made from ground observations with photo-interpretation maps have revealed that increasing complexity of geological deposits or increasing importance of properties not closely associated with landforms greatly reduces the accuracy. Also there is increasing awareness that the timing of air photography is critical. Photographs taken at inappropriate times may provide an unsatisfactory and even potentially misleading yield of information. Under European conditions photographs to record soil-tone patterns and soil change should be taken in winter and spring, the best months being March and April (Fig. 13.1). Crop patterns which reflect soil boundaries are best recorded in July and early August. These requirements must, however, be fitted into the period when conditions are suitable for aerial survey, so the planning of flights and placing of contracts should be carried out well in advance.

Suitable statistical methods for the testing of the goodness of soil boundaries drawn by photo-interpretation have become necessary following the increasing use of air photography. Multiple correlated attributes of the soil profile can be reduced to a single variate expressing a large proportion of available information by principal component analysis. This variate, the first principal component, can be plotted against distance on a linear transect and approximate positions of maximum (positive) and minimum (negative) slope found by inspection. These positions are then pinpointed using Student's t statistic. They represent the points where the soil boundaries – the maximum rates of change of soil with respect to distance – cross the transect. Beckett (1974) well reviewed the statistical assessment of resource surveys by airborne remote sensors, particularly in respect of the precision of information and the costs of obtaining data on soil conditions at different scales.

The use of colour photography as an aid in soil mapping only became important after 1960, following advances in technology by film manufacturers. Studies on the use of colour film for engineering soil and landslide studies showed that it gave a saving in interpretation time, often of about 50% in the USA. Research by British and Russian scientists indicated that colour aerial photography yields more data on soil conditions than panchromatic photography. Work has shown, however, that soil colours as recorded in the field and on aerial photography do not correspond exactly in respect of their Munsell colours; indeed overlap of hue, value and chroma is quite frequent. The influence of different soil hues on the optical densities of aerial colour and aerial infrared colour has subsequently been studied by densitometric methods. For example, statistical tests applied to density data obtained from one sample of 12 soils showed that the soils could be separated into two groups on the basis of significant differences in densities. One group consisted of those soils with *low chroma* (soils grey or neutral in colour) and which were best distinguished by infrared colour. The second group included soils with *high chroma* (intense soil colour) and which were best distinguished with colour film.

Infrared colour transparencies require suitable techniques for viewing and annotation and suitable light tables and filter frames were developed as more and more infrared colour was used. Techniques have also been developed for the production of panchromatic internegatives and positive prints from infrared colour film. There remains, however, the fact that transparencies are less convenient for use in manual photo-interpretation and this may partly explain the somewhat slow adoption of infrared by user agencies, despite its considerable information yield in respect of soil conditions and crop vigour. A summary of the advantages and disadvantages of different types of film used in aerial soil surveys is given in Chapter 4 (Table 4.2).

In order to communicate the photographic

Fig. 13.1 Soils on a former estuarine marsh and adjacent upland, near Martin, Lincolnshire: (a) under crops, July 1971; (b) bare soil, April 1971. Note the clear portrayal of the patterns of creeks that formerly traversed the marsh. The tonal variations in the photograph along the creek lines are primarily due to variations in soil texture and moisture content. (Source: Cambridge University Collection; Copyright reserved.)

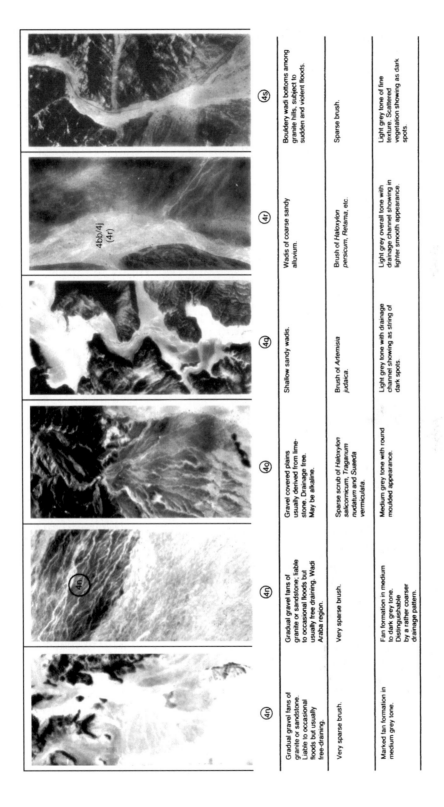

Fig. 13.2 A photographic key for soil and range conditions in Jordan. Some mapping categories, e.g. 4n, require more than one photograph to illustrate the appearance of a particular soil and range condition. This key was found to be effective when used in 1990, several years after its inception. (Courtesy, Aerofilms, Hunting Surveys Ltd, Boreham Wood, Herts.)

characteristics of soil and terrain units to other users, photographic keys of various kinds have been devised. Many are now available, including keys for vegetation, landforms, forest sites and soil conditions. An example of a photographic key for soil and range conditions in Jordan is shown in Fig. 13.2. The continuing interest in site selection is reflected by recent publications dealing with terrain evaluation, some of which are listed in the further reading recommended for this chapter.

13.4.2 MULTISPECTRAL PHOTOGRAPHY

Since soils are composed of mineral and organic constituents together with varying amounts of soil moisture, and as these constituents vary in amount and kind, it can be assumed that different soils will show different reflectance characteristics. There was, therefore, early work by soil scientists in the 1950s on the reflectance characteristics of different soil types, particularly in the infrared part of the spectrum. With the increasing use of multispectral aerial cameras and multispectral scanners in more recent years, there has been renewed interest in spectral signatures of surface objects, including soils. Recent laboratory measurements within the wavelengths of 0.32–1.0 µm have been made on soils ranging from wet (almost saturated) to dry (oven dried at 43°C) states.

From some 160 sets of curves three general shapes emerged (Fig. 13.3). Type 1 curves were characteristic of chernozem-like (mollisol) soils, type 2 of pedalfer (spodosol) soils, and type 3 of lateritic (oxisol) soils. An important part of the analysis of these results has shown that reflectance measurements made at only five wavelengths would suffice to predict with sufficient accuracy the other 30 wavelengths measured. It was recommended that the five selected wavelengths should be 0.4, 0.5, 0.64, 0.74 and 0.92 µm. Much more extensive studies of spectral signatures have now been made in the field as well as in the laboratory. Examples can be drawn from work in the USA on soils of

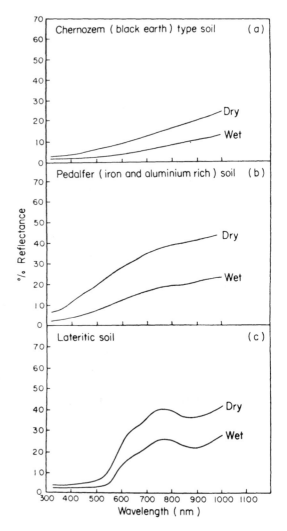

Fig. 13.3 Spectral characteristics of soils. (Source: Condit, 1970.)

Nevada. These spectra have revealed significant differences between saline and non-saline soils, and between gravel and silt loam soils in dry, wet and shaded conditions (Fig. 13.4). Other work in the UK has examined an extended spectrum of soils in the range 400–2500 mm (Fig. 13.5). Russian studies of the spectral reflectance of sands in the Karakum desert have shown that isopleths drawn by trend surface analysis of the spectral data can be used to distinguish three types of sand with differing mineralogical characteristics.

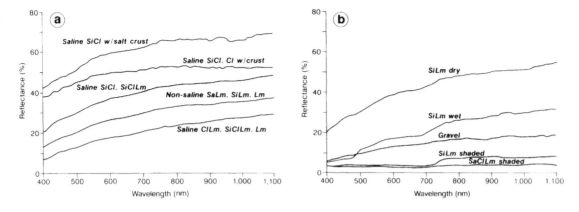

Fig. 13.4 Spectra of sunlit and shaded soils from ground-based studies in Nevada using a 15° spectro-radiometer. (Source: White and Henley, 1987.)

Spectral reflectance data have been examined also for the purpose of evaluating soil moisture and water use by plants through airborne remote sensing methods. Examples of correlation coefficients plotted against time for the available soil moisture in the sorghum versus film density for red filtered (8403 film) and infrared (2424 film) are shown in Fig. 13.6.

13.4.3 POLARIZED PHOTOGRAPHIC DATA

Although multispectral sensing for soil moisture assessment has received much attention, less work has been focused on the use of polarized visible light. When the unpolarized light waves are reflected from a soil surface, the degree of polarization at a high phase angle (Fig. 13.7) is much affected by surface soil moisture. Where free water stands at the

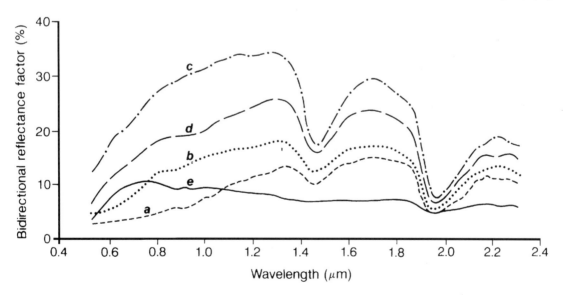

Fig. 13.5 Spectral reflectance curves for five UK soil types: (a) organic-dominated; (b) organic-affected; (c) minimally altered; (d) iron-affected; (e) and iron-dominated. (After Milton and Webb, 1984.)

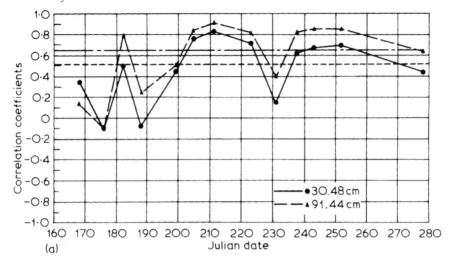

(a)

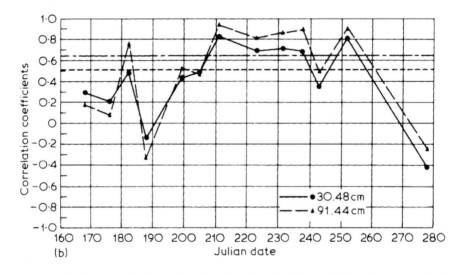

(b)

Fig. 13.6 Mission correlation coefficients plotted against time for the available soil moisture in the sorghum versus film density: (a) colour infrared film; (b) green-filtered black and white film. (Source: Werner *et al.*, 1973.)

surface, light will not be scattered but reflected specularly in one vibrational plane. For rough-surfaced soils below saturation the reflectance is diffuse and polarization low, but it will increase as saturation is approached. The intensity of polarized visible light reflected from the surface is related to a very thin surface layer of less than 1 cm. Laboratory studies have shown the relationships between soil moisture content, percentage soil moisture and phase angle (Fig. 13.8).

Early studies revealed that polarization of reflected sunlight provides the basis for a sensitive technique for remote sensing of soil surface conditions. In particular the polarization percentages have been shown to be related to both water content and soil type (Fig. 13.9). The technique seems particularly

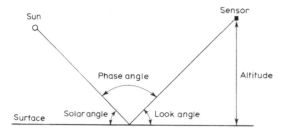

Fig. 13.7 Geometrical relationships and terms for the use of polarized photography. (Source: Curran, 1981.)

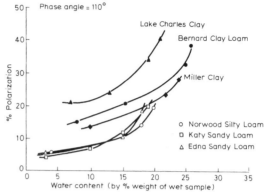

Fig. 13.9 Polarization related to water content and soil type. (Source: Stockhoff and Frost, 1972.)

useful when soils are at maximum moisture retention capacity, or near to this state. However, it should be noted that in some circumstances the degree of polarization is more dependent on soil type than on soil moisture and variations in soil aggregation are particularly important.

Some success has been obtained by analysing data obtained from a light aircraft using a camera equipped with a polarizing filter. Photographs were taken with the focal axis set to face the Sun and angled towards the ground surface. Two exposures were made of each scene, both through a polarizing filter on to a panchromatic film. The camera operator can obtain the record of maximum polarization by

turning the lines of the filter until they are first parallel with the ground surface and then at right angles to the surface. The densities of the imaged areas then can be measured quantitatively to show percentage polarization (Fig. 13.10) by using a densitometer, and standardization of densities for each scene are made by reference to the density of a marker such as a road or vehicle roof within the scene. Such a technique is, however, limited to very low altitudes (50–100 m) owing to atmospheric effects and to variability in soil areas sampled. Also, such factors as soil particle size, soil albedo and the range of moisture values will affect the correlation between polarized light and moisture content. In brief, the technique is best suited for obtaining an overview of moisture content at small sites where surface roughness is constant and data can be obtained in sunlit conditions.

13.5 REMOTE SENSING OF SOILS AND LANDFORMS BY NON-PHOTOGRAPHIC SYSTEMS

13.5.1 VISIBLE AND NEAR-INFRARED SCANNER DATA

Data from all the Landsat satellites have been used extensively in terrain studies. In

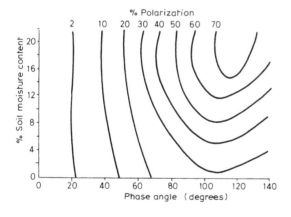

Fig. 13.8 The degree of polarization (PVL) recorded using a photometer–polarimeter in the laboratory for a range of soil moisture contents and phase angles. (Source: Steg and Frost, 1971.)

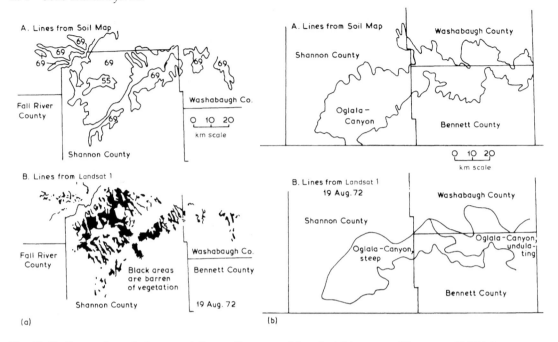

Fig. 13.13 Comparisons between existing soil maps and Landsat 1 imagery. (Courtesy, NASA.)

soil (Chapter 6) to assist in analyses of TM data. Mean reflectance values for Landsat TM Band 1 (blue 450–520 μm) and TM Band 2 (green 520–600 μm), TM Band 3 (red 630–690 μm), TM Band 4 (near infrared 760–900 μm) are the data commonly used. Band ratios then can be calculated for all TM Band combinations, e.g. through quantifying the so-called *Normalized Difference Vegetation Index* (NDVI: see also Chapter 16), using the following relationship:

$$NDVI = \frac{(\text{Band 4} - \text{Band 3})}{(\text{Band 4} + \text{Band 3})} \quad (13.1)$$

Studies in Nevada, USA, have shown that the NIR/Red ratio separates soils and vegetation into three groups: namely soil and senescent vegetation (ratios less than 2.3); grey and yellow green vegetation (ratios 2.0–7.5); and green vegetation (ratios greater than 7.5). It also has been found that NDVI separates the same classes with values of less than 0.3, 0.3–0.7 and greater than 0.7, respectively. Other simple band ratios can be effective

discriminants provided there is sufficient reflectance contrast between the two band passes, but the overall conclusion must be that such ratios cannot be used consistently for vegetation covers in all situations.

Special note should be taken of the initiatives of the US Geological Survey, the Soil Conservation Service and the Bureau of Land Management. In their work, Digital Elevation Model (DEM) data produced by the US Geological Survey together with Landsat MSS data were made available for all areas of continental USA. The agencies combined to

Table 13.5 Average probabilities of misclassification for a five-category recognition scheme (Yellowstone Park)

MSS Bands	Average probability of misclassification
5	0.086
5, 7	0.031
5, 7, 6	0.028
5, 7, 6, 4	0.027

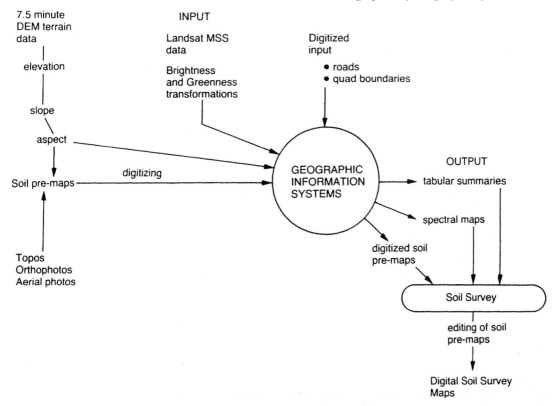

Fig. 13.14 Data processing flow for using Landsat and digital elevation data as inputs to soil surveys, USA. (Courtesy, US Geological Survey.)

develop analytical procedures for ongoing soil surveys. Slope class and aspect class maps were prepared to overlay with 7.5 min quadrangle topographic maps, and quads of corresponding orthophotography. The boundaries of slope classes were interpreted and adjusted with reference to interpretations from topographic maps, aspect maps and spectral class maps. Polygon maps were then created in the laboratory which represented the landscape as an estimate of soil distribution. The resulting map was described as a 'soil pre-map'.

The soil pre-map was then digitized and merged with spectral categories derived from classifications of geo-referenced Landsat MSS data and DEM derivatives. This merged data set was then used to create statistical summaries describing each polygon in terms of slope, aspect, elevation and spectral values. The

pre-map units so obtained were used to obtain delineations for the initial soil mapping units, and the statistical characteristics of each polygon provided quantitative data that have been used to develop preliminary definitions of landscape characteristics (Fig. 13.14). Initial soil taxonomy interpretations were developed and further refined through traditional field survey methods.

Landsat data have been used also by Ontario Geological Survey in Canada to map fuel-grade peat lands in that province. Landsat data recorded in mid- to late summer were geographically registered to 1:50 000 map formats. Following preliminary visual interpretation of wetland type, a selection of field sampling sites was made. Spotsampling in the field by means of helicopter transport allowed a wide range of *in situ* data to be collected, including peat depth, peat decomposition rating, pH of surface water, percentage cover by vegetation plant strata, and dominant vegetation. A supervised classification of the MSS data was then made. This led to the production of colourcoded maps of peatlands, with full geographic referencing. An area of 235 000 km^2 was mapped in this way. Accompanying reports summarize vegetation cover characteristics, peat depth and quality, and site characteristics. The total areas of fuel-quality and horticultural-quality peatland are estimated for each map sheet and a preliminary estimation of peat values was made.

The first generation of satellite sensors (Landsats 1–3) also have proved useful in mapping coarse geomorphic features, such as rivers, large streams, floodplains and terraces. In Arizona, USA, arroyos as narrow as 20–45 m wide have been detected on low-contrast and high-contrast Landsat images. The second generation of remote sensing satellites (SPOT and Thematic Mapper) have provided data suitable for much more detailed studies of geomorphic units and classification of soils at the soil series level. Simulated SPOT data obtained by a Daedalus AADS 1268 Digital Multispectral System, flown at 6.35 km altitude, have been used to study landscape components in New Mexico. The overall results of the study show that many specific features are well classified (>80%) but poor correlations were found for vegetation areas that are shrub-covered or contain fine-grained soils. However, these covered only 3% of the study area.

Second-generation satellites are now proving useful for computer-assisted approaches to drainage pattern classifications. The quantitative study of drainage basins and channel networks first developed by Strahler has led to the evolution of *structural pattern recognition*, which refers to digital image analysis methods that represent drainage patterns in terms of their intrinsic structure. The inherent characteristics of drainage patterns can be summarized as follows:

1. *Dendritic* patterns, with branching, tree-like systems, where tributaries join at an acute angle.
2. *Pinnate* patterns, resembling the dendritic but with secondary tributaries evenly and closely spaced, and parallel.
3. *Parallel* patterns, where tributaries flow nearly parallel to each other.
4. *Trellis* patterns, where primary tributaries are long, straight and parallel. Secondary tributaries are numerous and short.
5. *Rectangular* patterns, with clear right-angle bends in main stream and tributaries.
6. *Angular* patterns, with mixtures of acute, obtuse and right-angle junctions.
7. *Radial* patterns, with streams radiating from a centre like spokes of a wheel.

Thus, any drainage pattern consists of a main stream with primary and secondary tributaries (sometimes also tertiary), and different patterns have specific characteristics in terms of the forms and junctions of the component parts. The characteristics can be modelled using features such as shape, angle, elongation, bifurcation ratio and nodes. Computer software has been developed for analysis termed the Drainage Pattern Analysis (DPA) system which allows drainage network data from airborne or satellite sensors to be analysed.

13.5.2 THERMAL INFRARED IMAGERY

Although the scanners in the visible and near-infrared wavelengths are producing good results, there is considerable research into the applications of thermal infrared data for soil studies. The use of thermal infrared for soil temperature studies has been investigated in detail by NASA investigators using aircraft-borne sensors. Here, soil surveys were related to the thermal infrared imagery in a manner which suggested that surface soil temperatures can be indicative of subsurface soil conditions. Such results are to be expected in view of the effects of soil texture, composition,

Fig. 13.15 Infrared linescan imagery of estuarine lowland near Huntspill River north of Bridgwater. Note dark tones of water and clear representation of field drains by dark lines of lower temperature. (Courtesy, Royal Signals and Radar Establishment, Malvern; Crown copyright.)

porosity and moisture content on soil temperature regimes in different soils.

Thermal infrared scanning is particularly sensitive to moisture content at the very surfaces of soils. This is particularly the case when surface winds evaporate moisture from the surface layers. In these circumstances the cooling of the surface in exposed areas contrasts with the higher surface temperatures in areas protected from the wind. An example of these conditions is shown on Fig. 13.15, which shows part of an area of Somerset under conditions of 11°C air temperature and 4.5–7.9 m s^{-1} wind strength. This is an area of reclaimed estuarine clay that is under pasture and requires extensive field drainage. The open drainage ditches are readily distinguished by dark tones reflecting the lower temperatures of the water surfaces in relation to the land. The lines of buried field drains also can be detected as a result of the wetter soil conditions along these subsurface channels. The white-toned areas in the lees of the field boundaries reflect the shelter effects of the boundaries and the lower evaporation and

hence higher surface temperatures existing in these sheltered places.

The most concentrated space investigation of the relationships of soil temperatures to soil conditions has been through use of the Heat Capacity Mapping Mission, which was developed from the Applications Explorer Mission-A (AEM-A). The HCMM system provided for a circular sun-synchronous orbit at 600 km altitude. The major sensing system was the Heat Capacity Mapping Radiometer (HCMR), which included two bands (0.5–1.1 and 10.5–12.5 μm) and provided a resolution of 500 m with a swath width of 700 km. The HCMM satellite was launched on 26 April 1978. One of the principal investigations of the mission was made by European scientists through the 'Tellus' project, sponsored by the Joint Research Centre of the European Economic Community whose specific objective was to map diurnal soil temperature variations so that, by correlation with ground-truth observations, *thermal inertia models* could be developed for the prediction of variations in soil moisture content of the top soil layers. The

Table 13.6 Measured and estimated soil moisture and daily evaporation for Grendon Underwood, Bucks, England, 13 September 1977 (Source: Gurney, 1979)

Model	Evaporation (mm)		Soil moisture (%)	
	Estimated	Measured	Estimated	Measured
Bare soil (TELL-US)	0.85	0.62	29.6	30.1
Grassland (Tergra)	0.53	0.62	49.6	47.7

principle adopted was that soils of high water content would show lower diurnal variations in surface temperature than would drier soils.

Two digital models were used in the Tellus project. One model, the 'Tergra' model, was for use in grassland areas and used one measurement of the soil surface temperature. The second, the 'TELL-US' model, estimated daily evaporation and thermal inertia by measurement of both day- and night-time temperatures (Chapter 12). Estimated and calculated values for evaporation and soil moisture for a site in Buckinghamshire, England, are shown in Table 13.6.

13.5.3 MICROWAVE RADAR DATA

Knowledge of the spatial and temporal distributions of soil water is of economic and scientific significance through many agricultural, hydrological and meteorological applications. In view of the effects of soil moisture on the dielectric properties of surface materials, there is growing interest in microwave sensors for the detection of moisture states at the ground surface. Also the dynamic nature of water conditions in the landscape makes the microwave sensors attractive because they can be used on a regular basis, owing to their relative immunity from atmospheric effects, including cloud cover.

In summary, it can be suggested that the environmental scientist is mainly concerned with four aspects of soil moisture:

1. The *amount of moisture stored* at a given time. This can be expressed best on a volumetric basis or as a percentage of field capacity.
2. The *availability of soil moisture* to the higher

plants. The available water is determined by the water tensions occurring at the time of observation.
3. The *movement of water* into and within the soil. This movement is due to positive and negative pressures within the soil.
4. The *periodicity of waterlogging* in a given soil. The Soil Survey of Great Britain classifies soil into drainage categories on the basis of the relationships of soil profile morphology to periods of waterlogging (Table 13.7).

In order to use radar effectively for soil moisture determination, it is not only necessary that the backscatter coefficient, $\sigma°$, be dependent on soil moisture, but also that its dynamic range in response to soil moisture should be much larger than its dynamic range

Table 13.7 Soil moisture regime classes – wetness classes – and durations of wet states (Source: Curtis, 1977, after Soil Survey of England and Wales)

Class	
I	The soil profile is not wet within 70 cm depth for more than 30 days[a] in most years[b].
II	The soil profile is wet within 70 cm depth for 30–90 days in most years.
III	The soil profile is wet within 70 cm depth for 90–180 days in most years.
IV	The soil profile is wet within 70 cm depth for more than 180 days but not wet within 40 cm depth for more than 180 days in most years.
V	The soil profile is wet within 40 cm depth for more than 180 days and is usually wet within 70 cm for more than 335 days in most years.
VI	The soil profile is wet within 40 cm depth for more than 335 days in most years.

[a] The number of days specified is not necessarily a continuous period.
[b] 'In most years' is defined as more than 10 out of 20 years.

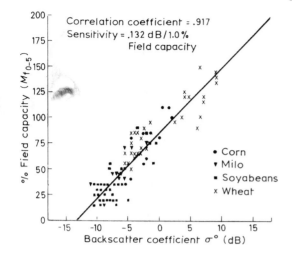

Fig. 13.16 Percentage field capacity in the 0–5 cm soil layer as a function of backscatter coefficient at 4.25 GHz, HH polarization, and 10° incidence angle, for corn, milo, soybean and wheat data sets combined. (Source: Ulaby, 1980.)

to other surface features, such as roughness and vegetation. Much of the initial work to determine the optimum choice of frequency, polarization and angle of incidence of the radar sensor was carried out by the University of Kansas. A Microwave Active Spectrometer was used from a Cherry-Picker platform to acquire a wide range of data beginning in 1972. The multiyear data, containing 190 data sets of bare fields, was used to generate an empirical model for back-scatter coefficient as a function of the volumetric moisture content normalized to percentage field capacity at 0.33 bar. This is illustrated by Fig. 13.16.

A model derived by Ulaby, one of the most successful researchers in this field, has been compared with data obtained using a C-band radar mounted on a crane boom over crops in France (Fig. 13.17).

Since the 1970s many further studies have been made by Ulaby and others of radar response in relation to soil moisture content. Most recently interest has centred upon the relationship between radar returns as a function of the percentage of field capacity. More experiments on different soil types are

necessary in order to decide whether percentage field capacity or soil water pressure (p_F) is the best soil parameter against which to make radar calibrations.

Important early studies of airborne and spaceborne investigations of radar response to soil moisture were undertaken in the USA by the University of Kansas and in Europe by means of the SAR 580 missions sponsored jointly by the Joint Research Centre of the EC and the European Space Agency.

The analysis of aircraft radar response to soil moisture made by the University of Kansas used a site south-east of Colby, Kansas, which afforded flat terrain with relatively uniform soils. It was instrumented with 39 raingauges and three meteorological stations and formed part of a site in the US Department of the Interior High Plains Experiment (HIPLEX). Three airborne radar sensors were operated at frequencies of 1.6, 4.75 and 13.3 GHz from a C-130 aircraft operated by NASA/JCS. The radars collected data at incidence angles of 5–60°. Dual polarization could not be obtained simultaneously. The study showed that the aircraft response to soil moisture is optimum at C-band frequencies and at angles of incidence of 10–20°. Like-polarization (HH) radar response was found to be by vegetation but in other studies has been found to be dependent on the character of row-tillage patterns (Fig. 13.17(a)). Cross-polarization (HV) data were also found to be unaffected by vegetation cover, but were also independent of tillage patterns (Fig. 13.17(b)).

13.5.4 FURTHER POSSIBILITIES

Interest in microwave methods continues to be fuelled by theoretical considerations which suggest that some *penetration of the soil* may be achieved thereby providing information in respect of all important subsurface moisture or soil layering. In principle this can be obtained by time domain measurements (Fig. 13.18). For a given range of soil relative dielectric constants and surface roughnesses,

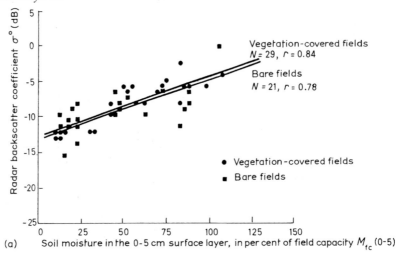

(a)

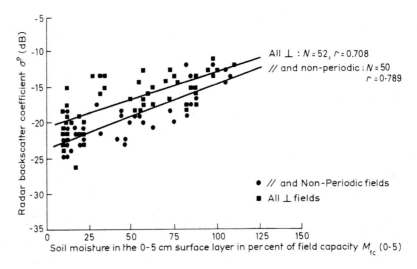

Fig. 13.17 (a) Comparison of radar soil moisture responses to bare and vegetation-covered fields. There is a minimum dependence of the response to vegetation at radar parameters of 4.75 GHz, HH polarization and 10° incidence angle. (b) Aircraft radar cross-polarization response to soil moisture for two categories of agricultural fields. Radar parameters are 4.75 GHz, HV polarization and 15° incidence angle. (Source: Bradley and Ulaby, 1980.)

approximately one-half of the energy of the incident wave will be reflected from the surface. Some portion of the wave energy that penetrates the surface layer will be reflected back to the observing platform from the first subsurface discontinuity or moisture accumulation gradient. The reflected energy from this interface is reduced with respect to the transmitted energy by the transmission loss through the surface layer and the three reflections the wave must undergo. In order to be able to detect the discontinuity, an extremely good range resolution would be required so as to detect the presence of the weak second reflection. This places a requirement on the radar system that the pulse duration be very

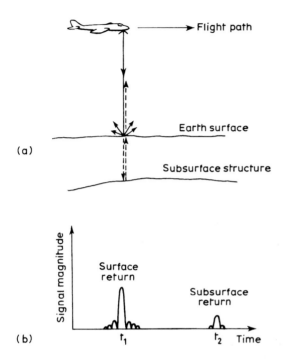

Fig. 13.18 Time domain measurement for subsurface features. (Source: Barringer Research Ltd, 1975.)

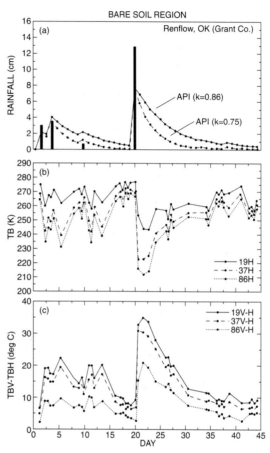

Fig. 13.19 Time series plots of (a) daily raingauge measurements (vertical bars) at Renfrow, Oklahoma and computed antecedent precipitation indices for two different recession coefficients, (b) brightness temperatures and (c) brightness temperature differences derived from SSM/I measurements. The measurements were over a region of sparse vegetation cover. (Source: Choudhury *et al.*, 1995.)

short, and makes the engineering tasks involved very severe. At the present time there is no evidence that systems are available to achieve this.

Interest in microwave methods is also growing because of the work on passive microwave satellite data which is at least beginning to yield real fruits (Fig. 13.19), and active microwave satellite data, which is at least as promising. It is clear, however, that information from active and passive microwave systems is complementary, a fact being recognized increasingly in new soil studies using remote sensing.

More generally, a wide range of new questions and problems have arisen in soil and landform research, many of which the instrument systems being developed for the EOS platforms are being designed to address. Until now most of the remote sensing research in these germane areas has been directed towards the improved recognition, evaluation and monitoring of aspects of soils and landforms, the investigations of which began long before satellites were first used for such purposes. The new questions that now demand attention from both conventional and remote sensing soil scientists alike include the following:

1. What and where are the unstable soil conditions that are subject to rapid changes produced by disturbance?
2. Are soils sources or sinks of carbon, nitrogen and other nutrients on a global scale?
3. What are the causes of desertification, and how can these processes be controlled?
4. What are the long-term effects of agriculture and irrigation on soil chemistry and productivity?
5. What is the time evolution of soils, and what can be learned about previous climatic conditions?

Meanwhile, increased attention will be paid to topography from satellites, both because this is interesting in its own right, and also because of its close relationship with soil development. More precise measurements of land elevation are providing an improved basis for studies of soil composition and height above sea-level; slopes determined by stereo-imaging of visible and SAR data are helping to provide better inputs to digital terrain models and models of surface run-off and landform development; and surface roughness measurements of unprecedented detail are assisting with estimations of surface maturity, soil abundance and particle size. As in so many areas – perhaps even all – of the natural sciences it is evident that the more we know of soil science, the more we still need to discover.

ROCKS AND MINERAL RESOURCES 14

14.1 THE USE OF AIR PHOTO-INTERPRETATION IN GEOLOGY

The use of remote sensing techniques in geology is long established through the use of aerial photography in photogeology studies. Sources of further information on the methods used in photogeology are cited in the bibliography at the end of the book.

Aerial photographs have provided a great deal of data for geological studies in all parts of the world. They can be interpreted to give information on the *structure* and *lithology* of rocks. In the study of structural geology such features as bedding, dip, foliation, folding, faulting and jointing can be observed. Aerial photographs provide evidence of bedding through the occurrence of ridges in the stereomodel, and differences in tonal response where beds differ in their mineral constituents. Sometimes certain beds can be recognized by a constant lithological interface which is so distinctive that they can be regarded as 'marker beds' or 'marker horizons'. The dip slopes of rocks can often be recognized more reliably on a stereomodel than on the ground because of the synoptic view of an area of dipping sediments obtained from the air. Accurate measurements of dip can be made by photogrammetric measurements on stereopairs. In

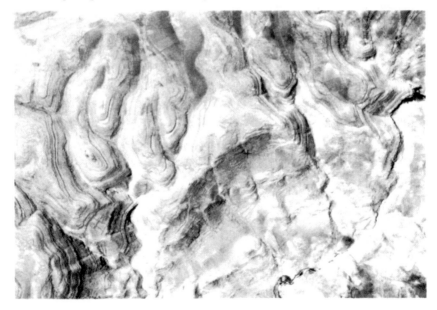

Fig. 14.1 An area of gently dipping sediments occurring in Jordan. Note how beds of differing lithology are characterized by different tones on the photograph. Field systems can be seen on the gentler slopes. (Courtesy, Hunting Surveys Ltd, Boreham Woods, Herts.)

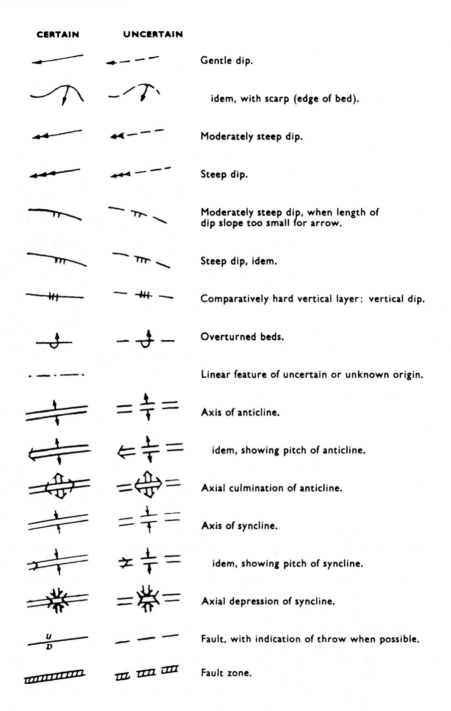

CERTAIN	UNCERTAIN	
		Gentle dip.
		idem, with scarp (edge of bed).
		Moderately steep dip.
		Steep dip.
		Moderately steep dip, when length of dip slope too small for arrow.
		Steep dip, idem.
		Comparatively hard vertical layer: vertical dip.
		Overturned beds.
		Linear feature of uncertain or unknown origin.
		Axis of anticline.
		idem, showing pitch of anticline.
		Axial culmination of anticline.
		Axis of syncline.
		idem, showing pitch of syncline.
		Axial depression of syncline.
		Fault, with indication of throw when possible.
		Fault zone.

Fig. 14.2 Selected symbols for photo-interpretation in geology. Where the identification of features is 'uncertain', the same symbols are used, but with broken lines as their bases. (Source: Shell Petroleum Company Ltd.)

Fig. 14.3 Oblique aerial photograph of the Fraser area, Canada (56°45′N, 63°35′W). Note the lineaments in the surface, which are made easily visible by the snow collecting in the lineament depressions. Both short and long lineament features are displayed. (Courtesy, Royal Canadian Air Force.)

regional mapping, however, geologists normally assess the dip by eye and place the dip into categories (e.g. <10°, 10–25°, 25–45°, >45°, <90°) (Figs 14.1 and 14.2).

Any discussion of *foliation* in rocks is made more difficult by the fact that the term foliation is used in different senses in Britain and America. British geologists distinguish between schistosity and foliation orientation so that, for them, segregation of such minerals into thin layers or folia is foliation, whereas parallel orientation of such minerals is referred to as schistosity. On the other hand, American geologists give the term foliation wider meaning so that lithological layering, preferred dimensional orientation of mineral grains and surfaces of physical discontinuity and fissility resulting from localized slip may be included in the term. In this chapter the American usage of the term will be adopted. Lineaments resulting

from foliation are generally parallel to one another but are short and do not normally consist of long continuous ridges or valleys (Fig. 14.3).

Folding in rocks is often apparent on air photographs where it would not be noticed in the field by a ground surveyor. The geologist can normally see a representation of the whole fold in the stereo model so that the axis and plunge of a fold structure can often be seen very readily (Fig. 14.4).

Faulting in rocks occurs where there has been a fracture along which rocks have slipped relative to one another. Generally faults provide fairly straight features and since the fracture is a zone of weathering and weakness, the surface manifestation is often one of negative relief. Where mineralization has taken place along a fault line, however, there may be positive rather than negative surface features. In most instances the most reliable evidence of faulting is displacement of bedding along negative linear surface features (Fig. 14.5).

Joints form patterns in rocks which are very similar in photographic appearance to faults, i.e. they often provide fairly straight negative features. The distinction between joints and faults can sometimes be made by careful observation to see if relative movement has taken place in the beds. If relative movement can be seen then the feature can be classified as a fault; conversely, if no movement can be detected it is better to record the feature as a joint. Jointing often plays a part in determining the patterns of river networks (Fig. 14.6). It also can be associated with topographic forms characteristic of granite areas and in such cases the boundaries of granite intrusions often can be plotted with some accuracy.

Whereas structural geology often can be interpreted with considerable certainty from aerial photographs, lithological interpretation is less easy. If the researcher is familiar with a particular field area and has used air-photos extensively it may be possible to recognize rock type with considerable confidence. However, where an attempt is made to recognize rock types in unfamiliar areas from photo features alone, the task becomes more difficult and uncertain. Various strands of evidence must then be used – in other words, the principle of 'convergence of evidence' must be applied. It has been suggested that the following stages may be included in interpretation of lithology and structure:

1. The recognition of the *climatic environment*, e.g. temperate, tropical rainforest, savannah, desert, etc.
2. The recognition of the *erosional environment*, e.g. very active, active, inactive.
3. The recognition and annotation of the *bedding traces of sediments or metasediments* (altered sediments).
4. The recognition and delineation of *areas of outcrop that do not indicate bedding*, e.g. intrusions in horizonally bedded rocks.

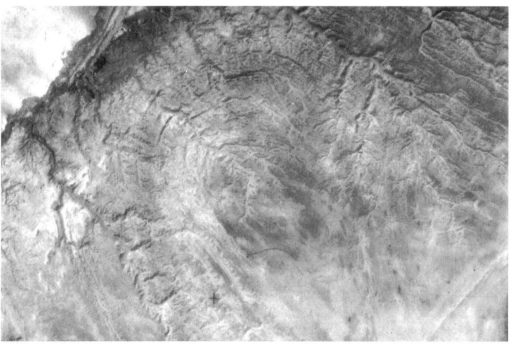

Fig. 14.4 A plunging fold structure near the Dead Sea, Jordan. (Courtesy, Hunting Surveys Ltd, Boreham Wood, Herts.)

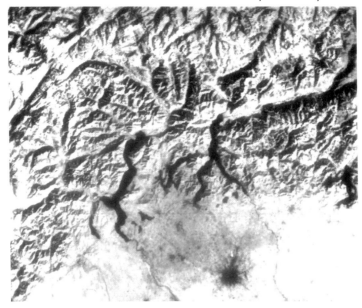

Fig. 14.5 Faulting in Alpine structures fringing the North Italian plain as seen in Landsat 1 imagery obtained in Band 7, 7 October 1972.

Fig. 14.6 Faulted terrain in an arid environment in Jordan. Note the sharp changes of direction in the wadi system resulting from the effects of faulting and jointing. (Courtesy, Hunting Surveys Ltd, Boreham Wood Herts.)

Fig. 14.8 Photogeological map prepared from a vertical photograph of folded sediments in Australia. (Source: Institute of Geological Sciences, London.)

k Sandstones forming prominent strike ridges

h″ Shale and siltstone with some sandstone

h′ Sandstone with some shale or siltstone

f Basal unit. Dark coloured surface, yielding pale coloured detritus. Lithology indeterminable

⌒ Unconformity

d Steeply dipping pale coloured sediments. Lithology indeterminable

Legend

—·—· Unit boundaries; certain, uncertain

⤳ Bedding trace and dip

— Fault

⟊ Fold axial plane trace; syncline, anticline

········· Unconformity

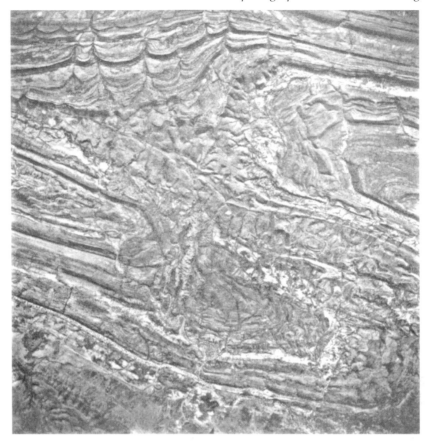

Fig. 14.9 Folded sediments in Australia. Note the pitching anticline on the left of the figure and the dipping sediments on the right. The photogeological map of this area is shown in Fig. 14.8. (Courtesy, E.A. Stephens, Institute of Geological Sciences, London.)

reduction in contrast of mid-infrared spectral features with increasing lichen cover, the spectral features of granites were still apparent with lichen coverage as high as 71%. Thus lichens do not appear to have a significantly deleterious effect on the thermal inertia estimated from spectral data. Nevertheless, further work needs to be done to investigate fully the usefulness of thermal inertia modelling as a basis for geological mapping.

14.2.2 MICROWAVE RADAR

Perhaps the most extensively used non-photographic system for geological studies is the microwave radar system. Side-looking radar systems are particularly valuable in equatorial and maritime regions where persistent cloud renders photography difficult to acquire. Furthermore the radar system illuminates the ground from the side so that shadowing effects are produced similar to those of low-sun photography (Fig. 15.9). These are well shown in the radar imagery of the Malvern Hills illustrated in Fig. 4.17.

Shape, pattern, tone and texture are used in the interpretation of radar imagery but the side illumination of radar systems places an emphasis on patterns, e.g. lineaments. Tones may vary greatly according to look angle and so must be used with caution. Textures in the imagery mainly reflect the physical form, e.g. roughness of the surface.

Lineaments are often well displayed in radar images (Fig. 12.9) and regional fault and fold patterns can be conspicuously displayed and readily mapped. The orientation of the radar system can, however, give rise to bias in the representation of linear features. Also, where areas are in the radar shadow (Fig. 4.17) ground information is lost.

14.3 SPACE OBSERVATIONS FOR THE STUDY OF ROCKS AND MINERAL RESOURCES

14.3.1 VISIBLE AND INFRARED IMAGING

The advent of satellite platforms has provided the geologist with a new and wider perspective on geological structures. Thus, for example, photographs of New Mexico obtained from the Apollo Saturn-6 spacecraft launched in April 1968 were examined with interest by geologists. They found a major north-east-trending lineament that was not shown on the most recent geological map available for the New Mexico area (Fig. 14.12(a)). However, the major impact of the satellite platform has been

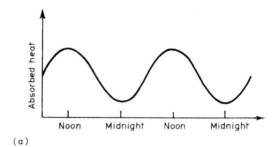

(a)

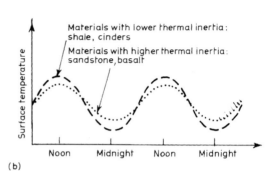

(b)

Fig. 14.11 Effect of differences in thermal inertia on surface temperature during diurnal solar cycles. (a) Solar heating cycle. (b) Variations in surface temperature. (Source: Sabins, 1978.)

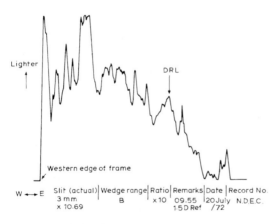

Fig. 14.10 Microdensitometer scan line of the infra-red linescan imagery of the Dugald River Lode, Australia. (Source: Custance, in Cole *et al.*, 1974.)

made by Landsat satellites with their MSS and TM data. Much geological work has been achieved with these types of data and some selected examples will be discussed below. The general achievements of the Landsat programme have resulted in keen interest on the part of geologists. The most significant results can be listed as follows:

1. The broadest use of Landsat imagery in geological mapping lies in the construction of *regional structural (or tectonic) maps*. Major geological features, contacts between distinctly different, thick rock units, and landform types can be effectively mapped from Landsat images with details comparable to and sometimes superior to mapping by conventional aerial photographic or ground methods.

2. *Computer-produced geological maps* at scales up to 1:25 000 can be made from Landsat-images with surprisingly good accuracy,

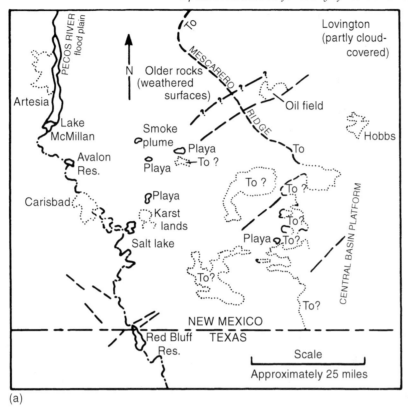

Fig. 14.12 (a) An example of linear features in the New Mexico area mapped from an Apollo photograph obtained in 1968. (Source: Carter and Meyer, 1969.)

provided training set data and other supervised methods are applied.

3. *Winter imagery*, with foliage-free scenes or snow-cover enhancements, provides useful data for mapping in many instances. Such imagery is especially valuable in seasonally vegetated areas and areas with considerable relief.

A number of applications for Earth resources satellite geological studies have been identified. In regions for which only poor quality or outdated maps have been produced, such data offers a method of constructing good general maps at small scales. Existing maps also can be checked against Landsat or SPOT imagery to correct mislocated or omitted rock unit contacts, geological structures

(fold axes, for example) and lava flows. Satellite-generated maps depicting structural information (particularly lineaments) are also useful in the search for ore deposits, oil accumulations and groundwater zones.

It must be borne in mind, however, that often field checking has been insufficient to verify the accuracy and correctness of satellite-derived data. Also it would appear that no reliable identifications of lithological types have been made consistently. The band ratio techniques (e.g. MSS 7/5) lead to some enhancement of visual images, which improves the separation of rock types. However, the ranges of ratio values for many common rocks and minerals overlap. Unique spectral signatures have not always been found to occur within MSS or TM sensing wavelengths. In fact the

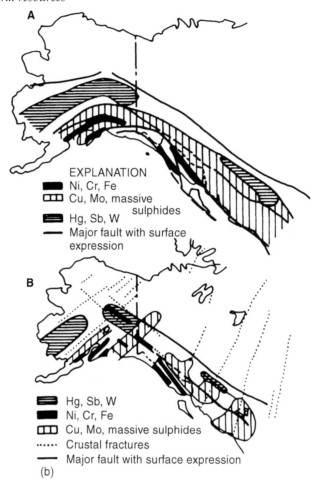

Fig. 14.12 (b) Geological conditions in Alaska interpreted from Nimbus 4 data: (A) conventional concept of the lithologic belts; (B) alternative lineations seen on the Nimbus image. (Source: Lathram *et al.*, 1973.)

ratio interval for haematite and serpentine – strikingly different minerals – is almost identical.

It also has been found that individual stratigraphic units are sometimes thinner than the linear resolution capabilities of the satellite. Thus remote sensing 'units' are often groupings of stratigraphic units and they sometimes have limited applicability to standard geological mapping procedures. With these advantages and limitations of satellite imagery in mind, one can examine some selected examples of geological interpretations in order to illustrate some of the techniques used.

An interesting early application of the mapping of lineaments and faults was provided by workers comparing Nimbus and Landsat imagery. Prior to the launch of Landsat 1 a cloud-free image of Alaska was obtained by the Nimbus 4 Image Dissector Camera System. This image showed a set of north-west and north-east trending lineaments which suggested previously unrecognized geological structures deep in the Earth's crust.

Lineaments and faults on Landsat images have been compared with known mineral deposits and fundamental fractures in the Canadian Cordillera. As a result it has been

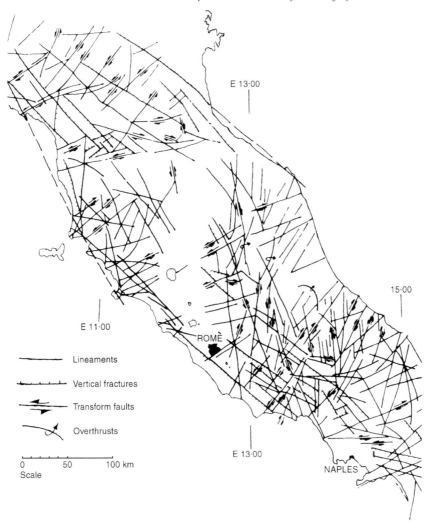

Fig. 14.13 Main fault systems in central Italy, as interpreted from Landsat data. (Source: Bodechtel *et al.*, 1974.)

possible to provide alternative maps to guide mineral resource application in Alaska and western Canada (Fig. 14.12b). The value of detail obtainable from Landsat images is also borne out by the detailed mapping of main fault systems revealed by Landsat data for the Italian peninsula, as shown in Fig. 14.13.

Many researchers have noted that snow cover does not obliterate lineaments; indeed, major structural features may be accentuated thereby. Observations in New England, USA,

indicate that a heavy blanket of snow (22.5 cm (9 in)) accentuates major structural features, whereas a light dusting of snow (2.5 cm (1 in)) accentuates more subtle topographic expressions. Comparisons of snow-free and snow-covered Landsat images for the same area have shown that snow-covered imagery allows more rapid fracture analysis and provides additional fracture detail.

Researchers have used the methods of spectral ratioing of reflected radiances of selected

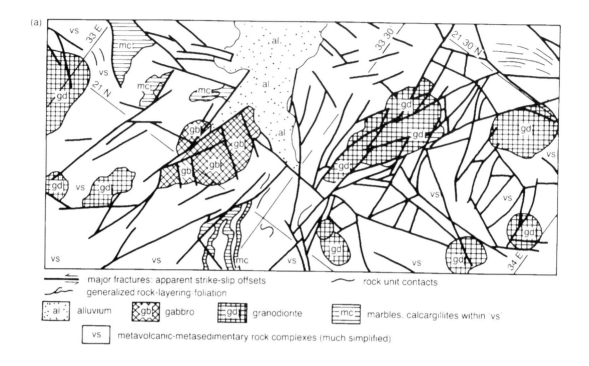

Fig. 14.15 Geological interpretation from (a) MSS imagery of study area in the Rocky Mountains (53 × 111 km) and (b) SIR-A imagery.

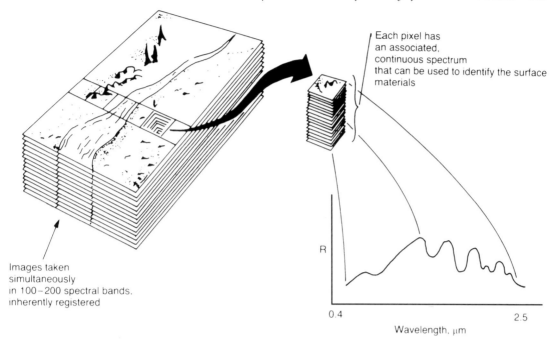

Each pixel has an associated, continuous spectrum that can be used to identify the surface materials

Images taken simultaneously in 100–200 spectral bands. inherently registered

R

0.4

2.5

Wavelength, µm

Fig. 14.16 The imaging spectroscopy concept. Two hundred or more images are acquired simultaneously, each in a narrow spectral band. In this manner, a complete reflectance spectrum can be constructed for every pixel in the scene. (Source: Vane and Goetz, 1988.)

data, have shown that satellite radar facilitates mapping of small-scale features owing to higher resolution and greater sensitivity to surface roughness. However, MSS or TM data do permit the mapping of rock units, owing to varying spectral response from rock surfaces, regolith, or surficial soils (Fig. 14.15). The advent of active microwave data (e.g. from ERS and Radarsat) has opened up many new opportunities for the mapping and analysis of geological features, especially in respect of features marked by significant variations in roughness at both large and small scales, including limestone surfaces, and volcanic landforms old and new.

14.3.3 TERRESTRIAL IMAGING SPECTROSCOPY

A new class of sensor called the imaging spectrometer is now under development for geological prospecting. The first images from an Airborne Imaging Spectrometer (AIS) were acquired in late 1982 and they constituted a new class of data requiring new approaches to information handling and extraction. The concept of imaging spectroscopy is shown in Fig. 14.16. Energy reflected from the Earth's surface is dispersed by a spectrometer and is used to form as many as 200 registered spectral images of the scene. The essential advantage of the technique is that each pixel in the scene has associated spectral information sufficient to allow for the reconstruction of a complete spectrum. Thus the diverse narrow-band spectral signatures that are characteristic of most surface materials can be used to identify those materials.

Field-acquired *reflectance spectra* (Chapter 6) for selected phenomena are shown in Fig. 14.17. Simultaneous imaging in many narrow contiguous spectral bands has required new thinking in respect of sensor design. Opto-mechanical systems, as operated from the

15.1 INTRODUCTION

There is now a growing awareness that *our use of nature's processes and resources must be adjusted to the limitations and requirements which it sets for us.* Thus in the twentieth century we have become more interested in the ecological changes which human settlements and industry have brought about in the local and the world environment. In particular we are now concerned with the *conservation of species* in danger of extinction and the *maintenance of ecological balance* in environments which have been altered to suit our economic needs. The important and fundamental fact remains that all animal life, including *Homo sapiens*, ultimately depends on plant life, which alone is able to synthesize elements into the form of food.

Wherever conditions are sufficiently favourable the climax vegetation cover consists of forest. In some areas, notably in savanna regions, forest clearance has led to an extension of grassland where forest formerly occurred. Approximately 42% of the total land area of the Earth is potentially forest land, 24% potentially grassland and 34% essentially desert.

Forests and woodlands are significant not only in their own right but also as balancing mechanisms in terms of CO_2 in the Earth's atmosphere. They are valuable in resource terms both to affluent societies in industrial nations and to poor rural communities in developing countries. Products such as timber, sawnwood and panels (for construction, doors, shuttering and furniture), pulpwood (for paper, cartons and rayon), poles, posts, mining timbers and railway track sleepers, fuelwood, fodder, fruits, pharmaceuticals, fibres, resins, gums, dyes, waxes and oils are some of the many items originating in woodlands. The value of annual world production of forest products exceeds US$150 000 million, and international trade is worth about US$60 000 million a year at present prices. Thirty countries (eight of them developing countries) each earn more than US$150 million a year (five more than US$1500 million) from forest products.

Alongside the continuing demand for forest products there is a huge burden placed upon forests and woodlands for fuel and as sites for shifting cultivation. More than 1500 million people in poor countries depend on wood for cooking and keeping warm. This annual consumption of wood is estimated to be more than 1000 million m³. In Africa the contribution of trees to total energy use is as high as 60%, and in South-East Asia and Latin America it is 42% and 20%, respectively. Around one fishing centre in the Sahel region of Africa, where the drying of 40 000 tonnes of fish consumes 130 000 tonnes of wood every year, deforestation extends as far away as 100 km. Fuelwood is now so scarce in the Gambia that gathering it takes 360 days per year per family.

There is now sufficient expertise available concerning forest and woodland management to plan for sustained cropping of woodlands whilst maintaining forest cover and habitats for woodland animals. Unfortunately, owing to over-use and exploitation of woodland areas the world position is now becoming sufficiently grave as to affect even world climates. In addition the habitats of many wild animals are under such pressure as to lead to the virtual extinction of many species.

Permanent pastures (land used for five years or more for herbaceous forage crops) and other grazing lands are found generally in areas unsuitable for crops without intensive capital investment. Their productivity is generally low, ranging from 3 to 5 grazing livestock units (GLU) per hectare in Central Europe to 1 GLU per 50–60 ha in Saudi Arabia. Even so, grazing lands and forage support most of the world's 3000 million head of domesticated grazing animals and hence most of the world's production of meat and milk.

Mismanagement of grazing lands is widespread. Overstocking has severely degraded grazing lands in Africa's Sahelian and Sudanian zones. In parts of North Africa, the Mediterranean and the Near East it contributes markedly to desertification. Overstocking together with uncontrolled grazing is also a serious problem in the Himalayas and Andes. Even in the upland areas of England and Wales there is evidence of overstocking affecting the natural composition of the rough grazings. In the world as a whole there are many areas where intensive grazing has removed both trees and grass cover, leaving the soil open to the risk of soil erosion.

Many of the extensive grazing areas of Europe are in upland areas of scenic beauty. In consequence they have been designated as 'protected areas' and efforts have been made to protect the natural vegetation by reducing the problem of overstocking.

Most of the critical problems in the management of grazing lands ('range resource management') and husbandry of grazing animals occur in remote areas where least is known about native grazing land. In most cases what is needed is better information on the following:

1. areas of usable grazing land by ecologically appropriate classes;
2. the ecological characteristics of each kind of range (phytosociology, plant succession, present range condition, autecology of the important species);
3. the special management problems associated with different grassland ranges;
4. indices of potential productivity of each kind of range

.

Information is needed also on the numbers and kinds of animals that make use of grassland resources. These include not only farm animals but also wild herbivores.

Wildlife is an important subsistence resource in developing countries, and a significant recreational resource in both developed and developing countries. In parts of Ghana, Zaire and other countries in west and central Africa, for example, up to three-quarters of the animal protein comes from wild animals. The nutritional importance of wild animals and plants for many people in developing countries is often underestimated or ignored.

Ecological studies require detailed biotic and abiotic information, some of which can be obtained only by ground survey, such as chemical data for the environment and organisms concerned. However, remote sensing has provided valuable support for ground sampling methods. It has proved particularly useful for *mapping plant associations* and *estimating the populations of larger animals* occupying remote regions.

In the case of plant ecology it may be noted that images from remote sensors are especially affected by variations in the growth forms of plants. Thus the broad vegetation categories of trees (Phanerophytes), shrubs (Chamaephytes) and perennial forms, such as the grasses (Hemicryptophytes), normally can be recognized readily. Also, the habitat of the community being studied can be viewed by examination of the images of the surface of the area. In this way the interrelationships of topography, local climate, soil, surface water, rock

Table 15.1 Wrongly classified points in a Landsat-based vegetation mapping project in the Värnamo area of southern Sweden, as confirmed by air-photo and ground survey data (Source: Bronge, 1995)

Vegetation class (abbreviated)	No. correct	No. wrong	Σ	% correct and 95% confidence interval
1.1 Old Spruce forest	40	0	40	100 ± 0
1.2 Young Spruce forest	32	6	38	84 ± 12
1.3 Pine forest	33	5	38	87 ± 11
1.4 Dry Pine forest	25	11	36	69 ± 15
2.1 Dry deciduous forest	30	6	36	83 ± 12
2.2 Moist deciduous forest	32	7	39	82 ± 12
2.3 Oak forest	11	23	34	32 ± 16
2.4 Beech forest	26	9	35	74 ± 14
3.1 Deciduous/Pine mixed forest	10	25	35	29 ± 15
3.2 Deciduous/Spruce mixed forest	27	8	35	77 ± 14
4. Clear-cuts	39	1	40	98 ± 4
5. Formerly open land (not evaluated)				
6.1 Meadow, dry-mesic	19	17	36	53 ± 16
6.2 Meadow, moist-wet	15	23	38	39 ± 16
7.1 Mire with coniferous forest	36	4	40	90 ± 9
7.2 Mire with deciduous forest	26	8	34	76 ± 14
7.3 Bog, hummocky type	38	1	39	97 ± 5
7.4 Bog, carpet type	28	12	40	70 ± 14
7.5 Bog, peat mud type	29	11	40	73 ± 14
7.6 Fen, lawn type	20	5	25	80 ± 16
7.7 Fen, carpet type (not evaluated)				
7.8 Fen, peat mud type (not evaluated)				
7.9 Fen, periodically flooded type	6	26	32	19 ± 14
8. Other areas (not evaluated)				

formations, human artefacts and vegetation cover can be assessed.

However, results of remote sensing surveys are never perfect, and knowledge of the error statistics for any study is both necessary, and valuable in helping to develop further improvements in monitoring strategies (Table 15.1).

Seasonal changes in plant forms usually produce marked differences in the images obtained by remote sensing. For example, winter and summer images of deciduous forest vary greatly from each other. The term *phenology* has been applied to that branch of science which studies periodic growth stages in the plant and animal world depending upon the climate of any locality. Phenological studies can be assisted greatly by remote sensing observations. Conversely, the interpretation of remote sensing data can be aided considerably by applying knowledge of local phenological changes to the study of images obtained at different times of the year.

Although only a minute proportion of the world's plants and animals have been investigated in respect of their *medicinal values*, medicine depends heavily upon them. It has been estimated that more than 40% of US prescriptions each year contain a drug of natural origin as sole active ingredient or as one of the main ones. The value of medicines from higher plants in the USA alone has been valued at US$ 3000 million a year, and rising. The most important applications of higher plants and animals for medicine are:

1. as constituents used directly as therapeutic agents (e.g. digitoxin, morphine, atropine);
2. as starting materials for drug synthesis (e.g adrenal cortex and other steroid hormones);

3. as models for drug synthesis (e.g. cocaine, which led to the development of modern local anaesthetics).

Genetic diversity and its preservation is an issue of great importance in respect of world survival. Genetic material presently contained in commercially important crops, trees, livestock, aquatic animals and organisms provides useful qualities for a limited time: they are rarely, if ever, permanent. New genetic material is often necessary in order to breed strains for improved yields, nutritional quality, flavour, durability, pest and disease resistance, responsiveness to different soils and climates, and other qualities. For example, the historic destruction of European vines by *Phylloxera* in the 1860s was overcome only when it was noted that the native American vine is tolerant of *Phylloxera*. Europe's wine production was saved by grafting European vines on to American rootstocks. This practice continues to the present day.

Even many useful breeds of livestock are at risk. Of the 200 or so cattle breeds in Europe and the Mediterranean, some three-quarters are threatened with extinction. The genetic pool contained in these rare breeds may yet be valuable. The very rare Wensleydale sheep was used to breed heat-tolerant sheep capable of producing good quality wool in subtropical lands. The Cornish hen, once a rarity, was crossed with other strains to produce the meatbird now used widely in the broiler industry.

The scale of the loss of domesticated plants and animals is matched in losses of wild species. Some 25 000 plant species and more than 1000 vertebrate species and subspecies are presently threatened with extinction. These figures do not take account of the great number of small species (e.g. molluscs, insects and corals) lost as their habitats are progressively destroyed. Some estimates suggest that from half a million to a million species will be made extinct by the end of the century.

15.2 REFLECTANCE FROM VEGETATED SURFACES

The reflectance of light from a vegetated ground surface is determined by several factors, which may include leaf geometry, morphology, plant physiology, plant chemistry, soil type, solar angle and climatic conditions. Plant structure, soil background and the surface condition of the plants' reproductive and vegetative parts are particularly important in determining spectral reflectance.

When considering reflectance it is, of course, necessary to bear in mind that the angular nature of reflectance can be described as either *hemispherical* or *directional* (Chapter 6). As we have seen, the term 'hemispherical' refers to circumstances where radiant flux is measured over a hemisphere, whereas 'directional' refers to collection in one direction only. It is convenient, however, to think of reflectance as being 'bidirectional', where the angles of both incidence and reflection are directional, as happens with satellite observations on a sunny day. We have discussed the bidirectional reflectance factor (BRF) in the context of field measurements of spectral response (Chapter 6). This is the ratio between the spectral radiance at a given angle of incidence and the reflectance of a standard reflector at the same angle within the scene. This measure is sometimes termed the *bidirectional reflectance* (BDR) for convenience. In remote sensing studies of vegetation the BDR relates to the vegetation canopy, consisting of a mosaic of leaves, and other plant structures, together with the soil background. Before considering the models used for the study of vegetation by remote sensing it will be useful to consider reflectance from a single leaf.

The characteristics of light reflectance and transmittance can be explained mainly on the basis of critical reflection of visible light at the cell wall–air interface of both the palisade and spongy layers of the mesophyll (Fig. 15.1). It has been shown that reflectance increases with an increase in the number of intercellular air spaces. This is because diffused light passes more often from highly refractive hydrated cell walls to lowly refractive intercellular air spaces.

The leaf structure is important in that the upper and lower epidermal layers, together with the palisade cells and spongy mesophyll, each play a part. Spongy mesophyll is important because it scatters near-infrared light. On the other hand, the lower sides of leaves reflect more light than the upper side and this is thought to be due to the absence of palisade cells on the lower side. Often the leaf becomes more spongy with age, with the result that the mature leaf displays less reflectance in the visible bands (about –5%) and more in the infrared (about +15%).

Frequently the different parts of the spectrum are affected differently by variations in plant composition and structure. The 0.5–0.75 μm band is characterized by absorption by pigments consisting mainly of chlorophylls *a* and *b*, carotenes and xanthophylls (Fig. 15.2). The 0.75–1.35 μm band is a region of high reflectance and low absorption which is greatly affected by the internal leaf structure.

The 1.35–2.5 μm band is influenced somewhat by internal structure but is more particularly affected by water concentration in the tissue. In general the spectral transmittance curves for mature and healthy leaves are similar to their spectral reflectance curves for the 0.5–2.5 μm bands but are slightly lower in magnitude (Fig. 15.3).

Leaf senescence occurs as the leaves end their functional life. Most herbaceous annual plants also have a progressive senescence from the older to younger leaves. During leaf senescence, starch, chlorophyll, protein and nucleic acid components are degraded. Thus the familiar autumn colours are caused partly by loss of green chlorophyll and the build-up of yellow and orange carotene and red anthocyanin pigments. Usually light reflectance increases markedly in the 0.55 μm (green) band when chlorophyll degradation takes place. Changes in leaf water content usually can be correlated with near-infrared reflectance but, in general, dehydration leads to increased reflectance over the whole range of wavelengths. Leaf senescence, however, leads to decreased infrared reflectance and the

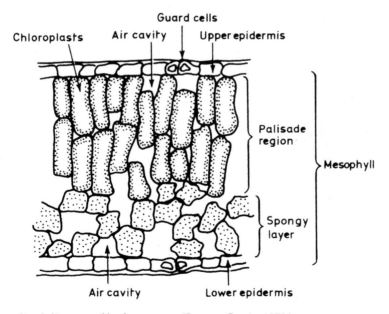

Fig. 15.1 A generalized diagram of leaf structure. (Source: Curtis, 1978.)

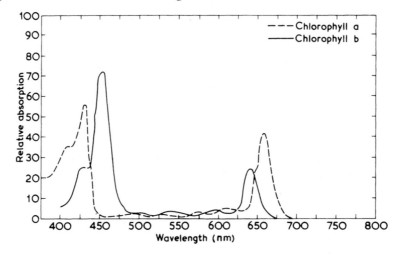

Fig. 15.2 Curves showing relative absorption of chlorophyll *a* and chlorophyll *b* as a function of wavelength. The combined absorption would be a summation of the two curves at each wavelength. (Source: Colwell, 1969.)

infrared plateau at about 0.75 µm is usually reduced considerably.

The reduction in infrared reflectance which characterizes senescence also occurs where disease strikes and the leaves lack their functional condition (Fig. 15.4). It is for this reason that infrared sensors have proved particularly valuable in studies of plant disease. It must be recognized, however, that the *resolution capability* of the sensor is important. Where there is a non-uniform distribution of dead and dying vegetation along with patches of more healthy vegetation, the classification becomes difficult if the pixel contains a mixture of elements.

Catalogues of the spectral properties of leaves have been derived mostly from laboratory determinations (Fig. 15.5) but the field spectral response may be rather different. As a result, analytical models of the natural scene have been developed and tested against real conditions. In many early studies the plant canopy was assumed to be an ideal diffusing medium having a uniform distribution of light scattering and absorbing components. More recent studies have envisaged a layered canopy, the biological components being represented by three flat 'Lambertian' plane sections. These planes are mutually orthogonal projections, one horizontal and two vertical.

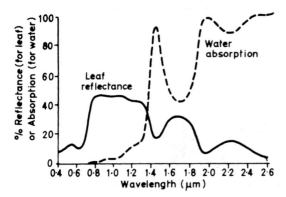

Fig. 15.3 A generalized spectral signature for leaf reflectance. Note the characteristic infrared 'plateau' at 0.75–1.2 µm and the regions of water absorption. (Source: Curtis, 1978.)

15.3 PHENOLOGICAL STUDIES

Models of the scattering and absorption by plant canopies requires estimates to be made of the leaf area affecting the response. Such estimates have been made by measuring the cumulative one-sided leaf area per unit ground area projected from the canopy top to a plane at a given distance above ground level.

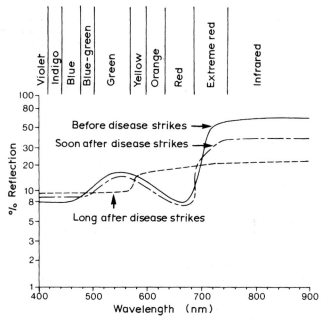

Fig. 15.4 Reflectance characteristics of healthy and diseased leaves. (Source: Colwell, 1969.)

Such measurements are termed the leaf area index (LAI) and can be measured for horizontal and vertical projections.

Vegetation has a characteristically high BDR variation between the red wavebands (strong absorption) and the near-infrared bands (strong reflectance). The relationship between a ratio of remotely sensed red and near-infrared BDR and measurement of LAI has been studied by various researchers (Table 15.2), initially using Landsat MSS data, but increasingly using NOAA AVHRR data also, as their suitability for lower spatial resolution but broader area and much more frequent vegetation assessments became appreciated (see also sections 13.5.1 and 16.3.3).

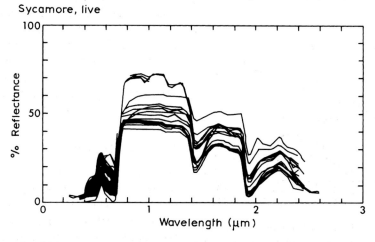

Fig. 15.5 An example of laboratory determinations of spectral reflectance curves. (Source: Leeman *et al.*, 1971.)

(a)

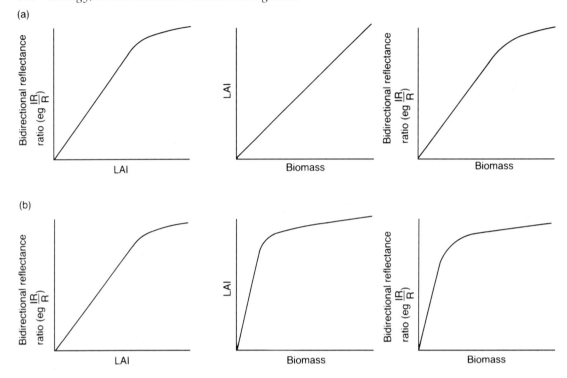

(b)

Fig. 15.6 The theoretical relationship between BDR, LAI and biomass for (a) a grass canopy and (b) a tree canopy. Note that while grass canopy biomass can be used as a surrogate for LAI, this is not so for the tree canopy.

It is also helpful to note that *any estimate of leaf area is likely to be indicative of the weight of plant material (biomass)* at the surface. The degree of correlation between BDR and biomass clearly will depend on the strength of the BDR/LAI relationships as well as the LAI/biomass relationships. For non-woody monocotyledons, such as the grass species, BDR is almost linearly related to biomass (Fig. 15.6). In other species (including woody species) that have high biomass (e.g. maize) the relationships are weak or inconsistent. Nevertheless species-specific studies can be made to establish BDR/biomass relationships. For example, measurements of *Calluna vulgaris* showed that there was a linear correlation in the low biomass growth phase. However, the growth of heather produces marked changes in the canopy morphology, with consequent changes in the relationships to reflectance over

time. Obviously such variations in response make estimates of biomass for mixed stands of heather much more difficult. Furthermore, although predicting that useful BDR/LAI/biomass relationships exist, the strength of the correlations with reflectance may be affected greatly by other factors, such as soil background, presence of senescent leaves, or phenological and man-made changes in the canopy.

In practice such LAI measurements can be correlated with plant stage and vegetation height (Fig. 6.5). Thus modelling of vegetation reflectance is closely concerned with the phenology of the plant types and the amount of biomass characteristic of different stages of growth. These relationships have been exploited in studies of the productivity of different ecosystems by remote sensing. For example, remote sensing data for shortgrass

Table 15.2 Seven of the more commonly used BDR ratios, where G = green, R = red and IR = near infrared

Name	Ratio
Simple subtraction	IR − R
Simple division	IR/R
Complex division	$\dfrac{IR}{R + \text{other wavebands}}$
Simple multiratio (vegetation index of normalized difference)	$\dfrac{IR - R}{IR + R}$
Complex multiratio (transformed vegetation index or normalized difference)	$\left(\dfrac{IR - R}{IR + R} + 0.5\right)^{1/2}$
Perpendicular vegetation index (vegetation directional reflectance departure from soil background)	$[(R_{soil} - R_{veg})^2 + (IR_{soil} - IR_{veg})^2]^{1/2}$
Green vegetation index (as above for use with Landsat satellite waveband)	$-0.29(G) - 0.56(R) + 0.60(IR) + 0.49(IR)$

prairie were used as inputs into biosystem models for the IBP (International Biological Programme) Grassland Biome programme in the USA.

The synoptic view provided by satellites can be used effectively in studies of the development of growth stages in vegetation over wide areas. For example, early NASA investigators described a phenology satellite experiment using Landsat data. In their study two phenological sequences were observed:

1. The *'Green Wave'*. A succession of images was used to record the geographical progression with time of foliage development (greening stage) as spring conditions extended over wide areas (Fig. 15.7).
2. The *'Brown Wave'*. A record of the geographical progression with time of vegetation senescence (maturation of crops, leaf coloration and leaf abscission). This progression plays the analogous role in the autumn to the Green Wave in spring.

Results from the Phenology Satellite Experiment confirm that it is possible to develop phenoclimatic models showing the progression of vegetational change during a particular year. For countries with highly developed agricultural or grazing economies, such information is very valuable. The LACIE programme (section 16.3.2) was used for determinations of crop status and crop yield and for management planning. Outside the cropped areas the ecological conditions of grassland ranges, particularly the steppe, prairie and savannah regions, are of great significance. The most important use of Landsat imagery for grassland ranges is that of monitoring changes in the forage conditions and development of grazing areas as the season progresses (i.e. Green/Brown Wave effects). It is then possible to determine areas and dates when plant growth ceases owing to drought or soil moisture depletion.

15.4 CONSERVATION IN NATIONAL PARKS

We can now turn to consider some examples of the application of remote sensing to conservation management. Interest in conservation has grown steadily as the effects of economic exploitation of animal, plant and land resources have become increasingly apparent throughout the world. In some measure this interest has been linked with growth in leisure time in the developed nations and the often expressed need for urban man to find places conservators is that measurements of changes in an ecosystem are time consuming, and often occur at a few points. Also, the sites where change takes place may be remote and expensive to reach. It is for these reasons that most conservators now include in their budgets some provision for air photography or other imagery. The savings in staff time, together with the value of a permanent record of transient conditions, make remote sensing an economic way of meeting managerial needs.

15.8 MARINE CONSERVATION

The role of satellite remote sensing in monitoring studies of oceanographic conditions is addressed also in Chapter 12. It may be noted here that the early studies using the Coastal Zone Color Scanner (CZCS) enabled scientists to evaluate the use of sensors to study a range of environmental parameters, including:

1. Formation of *yellow substance* ('Gelbstoff').
2. *Particulate materials* in the subsurface layers of the oceans (related to dynamics of phytoplankton). The concentration of suspended matter in the subsurface layers determines the penetration of sunlight in water. In this way suspended matter can be a limiting factor for phytoplankton production.

3. Marine *water quality*.

The pioneering CZCS sensor was launched in 1978 aboard the Nimbus 7 satellite. Since then, airborne oil surveillance systems have been developed based on aircraft sensor packages. Sensors used include a passive microwave imager (PMI), multispectral linescanner (IR/UV) with three channels (two infrared and one ultraviolet), side-looking radar and high-resolution aerial camera. The passive microwave imager has limited quantifying ability but provides some all-weather capability. Thin oil slicks have a low 'brightness temperature' whereas thicker oil slicks appear radiometrically warm. The multispectral scanner detects thermal differences between oil and water in IR bands, whereas UV detects relative differences between oil and water and will detect very thin slicks that may escape IR detection. Such systems can provide early warning of environmental changes that are damaging to sea organisms. The US Coastguard Aireye system is an example of such an airborne monitoring facility. An Italian system is described in more detail in Chapter 18. Further reading on aspects of remote sensing for studies in ecology, conservation and resource management is recommended in the bibliography for this chapter at the end of the book.

16.1 INTRODUCTION

The world population continues to grow apace (6000 billion is forecast for the turn of the century). It includes high populations in Third World countries that are undernourished. Yet, in advanced countries, such as Europe and the USA, there is now a problem of surplus production. One of the most pressing problems facing society is *how to feed a hungry world without causing permanent environmental damage* by intensive farming methods unsuited to the area concerned. In order to achieve this it is necessary to improve our knowledge concerning crop distributions and crop yields.

In the field of agriculture the main need is for early information on crop conditions and crop areas to allow for efficient management of the farming industry, and for planned distribution and marketing of produce. In an age when there is substantial (often ecologically damaging) government intervention in the form of support prices and subsidies there is a need for improved national and international data concerning:

1. crop production;
2. animal production;
3. farm structures (average size of fields, access, irrigation facilities, farm buildings).

Crop production is a function of *crop area × yield* (productivity per unit area). Thus there are three fundamental aspects to the production of information, namely crop identification, area measurement and yield prediction.

Furthermore, in order to be useful the data must be delivered to the authorities in a timely manner. For example, the area estimates in Europe are needed in December (winter cereals within plus or minus 4%) and May (spring cereals plus or minus 3%). Yield estimates for cereals are required in August. Thus early acreage determination of main crops, early yield predictions, information on plant diseases and pest attacks, quantification of the effects of moisture deficiency or waterlogging and continuous updating of inventories are some of the requirements for land use planning in modern agriculture.

At present agriculture dominates the land use of most countries, although forestry may play an important role in some. Nowadays agriculture is increasingly seen to be competing for land with the claims of recreation, mineral extraction, housing, industry and waste disposal sites as well as forestry. The Earth's surface consists partly of natural features such as vegetation, snow and ice (sometimes termed 'land cover') and features resulting from human activities ('land use'). Any inventory showing areas devoted to different purposes should be compared with the total land area concerned. Therefore, care must be taken to account for land cover areas (e.g. streams, moorland, etc.) even though the main purpose may be to establish areas of a particular land use, such as crops. A land use and land cover classification suitable for use with remotely sensed data is shown in Table 16.1. Information relevant to many of the 'land cover' categories can be found in sections of the book dealing with water, ecological studies, resource management and the built environment.

Agricultural support has generally absorbed some 70% of the European

techniques in mapping agricultural land use from air-photos is given in the *Manual of Photographic Interpretation* first published by the American Society for Photogrammetry in 1960. At this time features such as patterns of fields, tone, texture, shape and size of the cropped areas were used as aids to interpretation. Further evidence was sought where photos were available for different seasons. Changes in the photographic appearance of particular areas often could be related to conditions of tillage, crop growth, crop rotation or conservation measures. As in all manual interpretation work the best results were obtained by a combination of field knowledge and careful examination of the images. The same principles apply today in the visual interpretation of both aerial photography and satellite pictures.

At a very early stage American and British research workers recognized that aerial photography was useful for the detection and mapping of crop diseases and pest infestations. However, the development of colour, infrared colour and multispectral photography has added further impetus to this area of research in the last decade. Diseases including potato blight, take-all disease in cereals, leaf spot in sugar beet and yellow dwarf disease in barley have been observed and their effects on crop conditions noted. Likewise the effects of nematode attacks (eelworm) and soil fungi may be detected and affect the spectral response of a particular crop (Fig. 16.2).

Where crops have been affected by disease or adverse environmental conditions, dead portions of the crop can be identified readily by using infrared colour photography. On this emulsion, healthy plants image in red whereas diseased and dead portions assume different colours. The colours of diseased plants often range from salmon pink to dark brown, depending on the severity of the attack. Dead portions image in green or bluish grey.

Crop discrimination using infrared colour photography has been studied closely and it has been found that the percentage accuracy depends on time of year, location and environment. For example, in north-eastern Kansas in late July all crops were found to image red. In Britain, however, it is noted that the quality of grass pasture often can be detected by variations in the intensity of the red hue, and in some cases this is a reflection of the fertilizer treatment the field has received (Plate 11, colour section).

16.3 MULTISPECTRAL SENSING OF CROPS

16.3.1 BASIC RESEARCH

In view of the different spectral responses observed for different plant types, a great deal of effort has been made to measure the spectral reflectance characteristics of a number of crops. A major source of data concerning crop signatures was obtained under the Air Force Target Signatures Measurement Program in the USA. These were mostly measurements in the visible wavelengths, and much of the data was derived from laboratory investigations. Examples of the signatures obtained for certain crops are given in Fig. 16.3. The spectral response of a field crop depends partly, however, on the layering within the crop and so modelling of crop canopies also is important (Chapter 15).

The principal objective now being pursued is that of automatic recognition of land use categories and crops by means of computer analyses of spectral reflectance data. Such analyses can be applied to either airborne sensor data or satellite data.

There are broadly two approaches which may be adopted in mapping from spectral data. On the one hand a 'training set' of data can be obtained from matching known crops to sets of data from particular sensor systems. This training set then can be entered into the computer so that the computer classifies all similar data as the crop in question. Such classifications are termed *supervised* classifications. Alternatively, it is possible to adopt an *unsupervised* classification approach. In this case the remote sensing data are grouped and

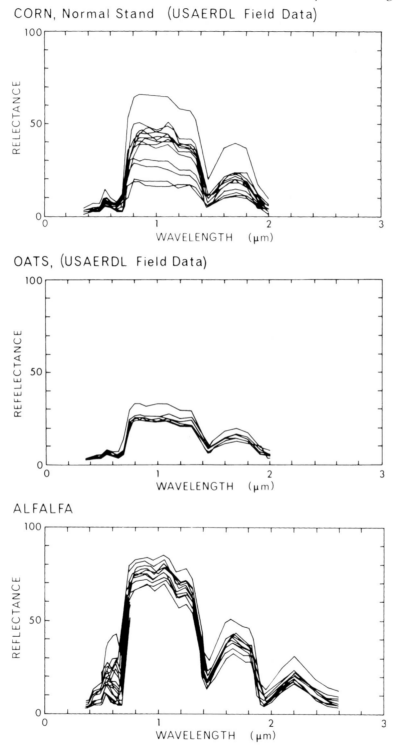

Fig. 16.3 Spectral signatures of corn, oats and alfalfa. (Source: Leeman *et al.*, 1971.)

field observations, photo-interpretation and automatic analysis of MSS data. Five categories of corn blight severity were used in the field. The best correlations were achieved when the blight symptoms were well established. It also appeared that the correlation between MSS analysis and field data became better at later dates in the study, thus emphasizing the importance of phenological conditions in such studies (Fig. 16.6). This important early study concluded that neither manual interpretation of small-scale photography nor computer-made analysis of MSS data gave adequate detection of corn blight during early stages of infection. Analysis of the data did, however, permit the detection of outbreaks of moderate to severe infection levels.

16.3.2 INTEGRATED STUDIES IN CROP MONITORING USING MULTISPECTRAL DATA

It is instructive to review developments in the USA over the past three decades towards a crop monitoring system. The Agricultural Board of the National Research Council set up its Committee on Remote Sensing for Agricultural Purposes in 1961. In the 1960s the multispectral concept was developed and initial feasibility experiments were carried out. Improved airborne multispectral sensors were developed in the mid-1960s and the development of decision algorithms to analyse multispectral data soon followed. Thereafter there was a long period (1966–1972) when controlled field experiments were carried out, together with over-flights with an airborne scanner. Then in 1969 a Landsat 1 multispectral scanner was flown on Apollo 9. Subsequently the Corn Blight Watch and the launch of Landsat 1 in 1972 gave added impetus to both analytical techniques and concepts. As a result, in 1974 the pioneering LACIE (Large Area Crop Inventory Experiment) programme was initiated to test the technology and know-how developed by applying it to the assessment of crop production over several of the most important agricultural regions of the world (Fig. 16.7). The objectives of the LACIE experiment were:

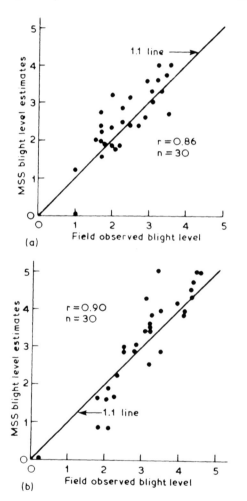

1. To evaluate the applicability of space technology to the global monitoring of agricultural crops.
2. To integrate the necessary components into an experimental system to permit the acquisition and analysis of the required volume of data in a timely manner.
3. To conduct the experiment in a quasi-

Fig. 16.6 Correlation of field observation and machine-assisted analysis of multispectral scanner data estimates of segment average blight severity levels, 23 August (top) and 6 September (bottom). (Source: MacDonald *et al.*, 1973.)

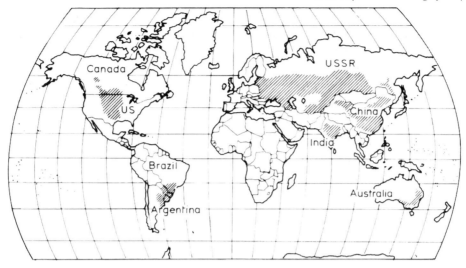

Fig. 16.7 Large Area Crop Inventory Experiment (LACIE) study areas.

operational manner to evaluate the technology under conditions representative of any
future operational missions.

4. To assess the suitability of the 'first-
generation' technology and identify areas
requiring improvement and suggested
means of development.

5. To meet an accuracy goal such that the
estimates at harvest should, on average, be

within +10% of the true country production
90% of the time.

6. To set a timeliness goal such that Landsat
data should be reduced to acreage information within 14 days after acquisition.

The LACIE study was focused on monitoring wheat production in the selected regions
shown in Fig. 16.7 and the LACIE technology

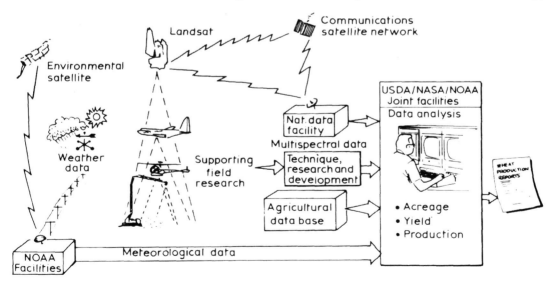

Fig. 16.8 The LACIE system data flow.

inaugurated the AgRISTARS (Agriculture and Resources Inventory Surveys Through Aerospace Remote Sensing) project in 1980. Its primary purpose was to assess the feasibility of integrating aircraft and satellite data into the data collection systems used by the US Department of Agriculture (USDA). The programme led to much progress, especially in the use of NOAA and geostationary environmental satellite data in support of crop monitoring, particularly in respect of meteorological data useful for crop forecasting models. Unfortunately budgetary problems had reduced the programme substantially by the mid-1980s, and although TM data were available to AgRISTARS investigators after 1982, these data, supporting improved crop recognition, were never as well integrated as those from LACIE. Today, even with both TM and SPOT data available, Morrison's view still holds good, especially in areas of small fields and mixed crops of similar types.

16.3.3 REMAINING PROBLEMS IN MULTISPECTRAL SENSING

Europe is one area of particular difficulty, where the problem of field size is reflected in the small size of the agricultural holdings (Table 16.3). It will be evident that about 46% of holdings are less than 20 ha in size. In these small holdings several crops often divide the farm into parcels smaller than 1 ha. It can be estimated that correct recognition of a 0.5 ha parcel would require a pixel size of no more

Table 16.3 Number and area of agricultural holdings with 1 ha arable area and over by size groups in the EC (Source: Eurosat)

Size group	Quantity	Area (ha)
1–5 hectare	2.429×10^6	6.14×10^6
5–10 hectare	1.067×10^6	7.67×10^6
10–20 hectare	1.067×10^6	15.14×10^6
20–50 hectare	0.845×10^6	25.32×10^6
>50 hectare	0.2961×10^6	34.48×10^6

The % of the holdings with ≥20 hectare is 54%

than 25 × 25 m. Furthermore, in Europe the cloud cover makes it difficult to obtain more than two repetitive scenes using visible sensors. Present experience indicates that to achieve an accuracy of 95% in acreage estimation, using only two multispectral scenes, would require a ground resolution of 20 m.

It is also being recognized increasingly that the problem of modelling the radiance of crop canopy is, in many ways, more complex than that of modelling a complete cover of natural vegetation. As the crop grows, different plant characteristics are developed and simultaneously the amount of underlying soil visible to the sensor changes through the growing season. As we have noted, remote sensing devices detect radiance primarily from the canopy but additional complicating factors must be borne in mind. The reflectance characteristics are dependent on a number of variables which change during the growing season, including:

1. optical properties of stems and reproductive structures within the canopy;
2. leaf area indices (LAI);
3. canopy densities;
4. leaf orientations and shapes;
5. foliage height distributions;
6. transmittance of canopy components;
7. the amounts of soil background viewed and their reflectance contribution.

With regard to optical properties we have noted already (Chapter 14), the relationships between chlorophyll and the red/infrared portions of the spectrum. As chlorophyll changes with maturation of the crop so does the *wavelength of the reflectance*. The sharp change in leaf reflectance over the wavelengths 0.68–0.75 μm has been called the 'red edge'. Experiments have shown that chlorophyll content, background material returns, multiple leaf stacking and leaf water content affect the maximum slope of the 'red edge'. The following conclusions have been made:

1. Shifts in the wavelength of the 'red edge'

are related directly to leaf chlorophyll content. An increase in chlorophyll is marked by a shift of the 'red edge' to *longer wavelengths*.

2. Provided the 'red edge' is measurable it can be used to distinguish live vegetation from other background targets.

3. 'Red edge' measurements over sufficiently narrow wavelength intervals can be used as early indicators of *crop stress*.

For crop canopy studies based on MSS data it has been found useful (as explained in Chapters 5, 13 and 15) to use normalized difference vegetation index (NDVI) values relating MSS band 5 (red) and MSS band 7 (infrared) data, i.e:

$$NDVI = \frac{IR - R}{IR + R} \qquad (16.1)$$

Early in the modelling of crop canopies by such a method the *brightness* and *greenness* of a crop were tackled by seeking *orthogonal indices* using a linear function derived from four Landsat MSS bands. The new axes that were developed comprised what was termed the 'Tasselled Cap'. This transformation rotated the MSS data so that the majority of the information was contained in two components of the physical scene, i.e. brightness and greenness.

Subsequently the Tasselled Cap concept was extended to six-band Thematic Mapper data. In this case the rotation was adjusted so that three dimensions were defined which broadly accorded with soils, vegetation, and a transitional zone that might be associated with conditions of canopy wetness and/or soil moisture.

Meanwhile other vegetation index models were developing, including the perpendicular vegetation index (PVI). We have already noted that as vegetation grows the red reflectance decreases and the near-infrared increases. The perpendicular (or orthogonal) distance of a vegetation spectral plot to the soil line is, therefore, a measure of greenness. The perpendicular vegetation index defines for each pixel the distance that the crop reflectance is from the plane of soil reflectance in the feature space defined by the near-infrared and visible wavebands (the 'soil line'). One may calculate PVI as follows:

$$PVI = [(R_s - R_v)^2 + (NIR_s - NIR_v)^2]^{1/2} \quad (16.2)$$

where R is the red waveband reflectance, NIR the near-infrared waveband reflectance, and s and v the soil and vegetation respectively. A positive PVI indicates vegetation, zero indictates bare ground and a negative value indicates water (see also Table 15.2).

By the time of writing, as many as about 50 different vegetation indices have been devised, all of which have been said to show better sensitivity than individual spectral bands for green vegetation detection. The earlier indices were purely statistical, whereas more recently developed indices have sought to be more physical and 'explanatory'. The great range and variety of vegetation indices has arisen because the satellite measurements of plants are affected by many variables – including the atmosphere; sensor calibration and viewing conditions; solar illumination geometry; soil moisture; vegetation and soil colour; vegetation type, condition and behaviour; and, therefore, target brightness. In conclusion, it has been rightly remarked that: 'For whatever application, the choice of a vegetation index is quite delicate to make.'

Turning from spatial to temporal variations in vegetation and crop appearance, the work of the Joint Research Centre, Ispra, has shown that *monotemporal* crop classifications of Landsat 5 TM data contain a high level of spectral confusion. Furthermore, it has been found that *multitemporal* data do not always provide sufficient information for classification. However, Tasselled Cap transformations have been found to present efficient compromises in many cases, although precise geometric correction has been found to be necessary.

At present the task of modelling crops effectively remains formidable, although considerable progress has been made in the generation

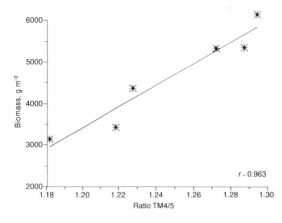

Fig. 16.11 Functional relationship between TM ratio (4/5) and biomass of corn based on regression analysis of data from 27 June, 20 July and 14 August 1986, Freiburg, Germany. (Source: Mauser, 1988.)

of adjusted soil-line data. Recent results are tantalizingly near to achieving operational accuracy under favourable circumstance but there is a need to use sensors with 20 m ground resolution with visible and infrared data. However, it has also been shown that land cover/crop classifications are affected by the sensor view angle. In general, highest classification accuracies are obtained at view angles close to those at which the training sample data were obtained. Thus, care needs to be taken in to the use of SPOT data which are obtained at different viewing angles. Summarizing the position so far as the use of Landsat and SPOT are concerned in crop analysis, the following conclusions may be drawn:

1. The *crop calendar* is a very important guide as to which data sets are likely to be most useful and should serve as a basis for ordering data.
2. *Multitemporal image analysis* is likely to improve classification accuracy compared with monotemporal studies.
3. *Image ratios and transformations* of data are useful in improving crop classification accuracy. In addition they can remove redundant data.
4. *Field size, topography and variability of crops*

within a field can have a marked effect on classification accuracy.
5. SPOT data are improved by the addition of a *mid-infrared channel* but the user must be aware that the different viewing angles obtained by SPOT can have an impact on classification accuracies.

The importance of including infrared data in crop studies has been borne out by many studies. Thematic Mapper data have presented advantages in this respect. For example, studies in the Upper Rhine near Freiburg, which were carried out in relation to the AGRISAR radar campaign of 1986, used TM data for estimates of biomass. Promising functional relationships between multitemporal ratioed TM bands 4 and 5 and the recorded biomass of corn were obtained (Fig. 16.11).

In conclusion, the growing trend towards multiparameter 'knowledge-based systems' should be noted, in which data such as altitude, temperature, and existing map information can be used in the classification stage, plus the increasing interest in 'all-weather systems' such as radar for those areas where multitemporal data are difficult to obtain because of frequent or persistent cloud cover.

16.4 RADAR SENSING OF CROPS

Two aspects of crop quality lend themselves to radar monitoring: *moisture status* and the *topology of the crop*. A strong correlation between the radar back-scattering coefficient and plant moisture content was found by workers at the University of Kansas using a 9.4 GHz radar system mounted on a platform 26 m above the ground. Linear regression analyses of the radar back-scattering coefficient ($\sigma°$) on plant moisture provided good correlation of about 0.9 and a slope of −0.275 dB per cent plant moisture with nadir observation. Also it was found that $\sigma°$ undergoes rapid variations shortly before and after wheat is harvested (Fig. 16.12). These variations suggest that radar could be used for estimating wheat

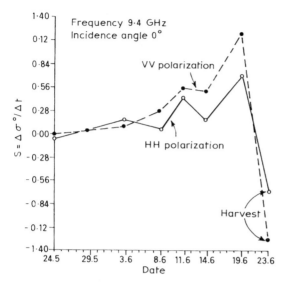

Fig. 16.12 Changes in radar response as a wheat crop matures in Kansas. The period of observation extends from 24 May to 23 June 1974. (Source: Bush and Ulaby, 1975.)

maturity and for monitoring the progress of harvest.

In parts of Europe there may be only a few days in the year when conditions are ideal for Landsat observations. In these circumstances the use of radar has many attractions. In cloudy regions throughout the world it may be the only sensor capable of providing data for time-discriminant analysis. Evaluation of radar has taken place in the USA, Canada and Europe in the last two decades and the next decade will see continued emphasis on radar studies. Perhaps the most promising avenues of research have been concerned with ratioing polarized radar returns at different grazing angles and using the data in time-discriminant analysis (Fig. 16.13).

Some of this work of an experimental nature was carried out under the auspices of the European AGRISAR campaign of 1986. This obtained SAR images over various sites using the French VARAN-S system mounted on a B-17 aircraft. The campaign, which was organized by JRC, Ispra, aimed to provide multipolarized X-band SAR data for various agricultural regions. Temporal values for backscatter were obtained from different cereals. An example from a test site in the Upper Rhine near Freiburg shows that winter and summer cereals can be distinguished. The X-band SAR images that were used were of different polarizations and dates: 27 June (VV), 17 July (HH) and 13 August (VV). For cereals, separation was best in June and July (the crops had been harvested by the August date). Other crops studies (Fig. 16.14) were best separated by means of July data. Although present SAR technology and analytical methods are far behind those used for optical data there are

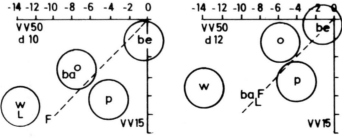

be beets, p potatoes, w wheat, o oats, ba barley, L *Lolium*, F *Festuca*

Fig. 16.13 An example of the discrimination of different crops by a ground-based X-band radar observing at different angles of incidence (VV 50°: VV 15° on day 10 (d10) and day 12 (d12)) of a series of observations. The circles are drawn at 2 decibels radius and VV indicates vertical polarization. (After Kasteren and Smit, 1977.)

indications that crop types can be distinguished using multitemporal X-band data. The analytical processes are lengthy and complex so that an operational system seems to be some distance away.

16.5 CROP YIELD MODELLING USING INFRARED SENSING

Whilst thermal infrared sensors operating from aircraft can provide high-resolution imagery capable of distinguishing hedges, trees and animals (Fig. 4.10), equivalent satellite sensors have more limited ground

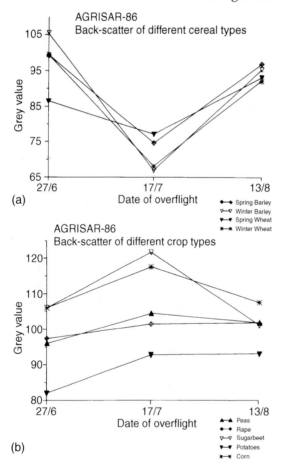

Fig. 16.14 Grey value profile of the AGRISAR data for (a) different cereals in the test site, and (b) non-cereal cover types. Grey values are in arbitrary power units. (After Mauser, 1988.)

resolution. The Ground Resolution Element (GRE) of the Heat Capacity Mapping Mission (HCMM) was about 500 m, whereas Band 6 (10.40–12.50 μm) on Landsats 4 and 5 provided a GRE of 120 μm. As a result of this, low-resolution thermal infrared data are of little use in crop identification. Such data become important, however, when used in crop yield (productivity per unit area) models.

The two primary environmental determinants of crop yield are *temperature* and *moisture*. The technology required for remote measurement of surface temperatures of soil and crops is well developed. Thus the crucial issue for the production of crop yields by remote sensing data remains the assessment of available soil moisture (see also Chapter 13). Temperature measurements may, however, provide valuable information and there have been efforts to model the relationships between air temperature, leaf temperature and crop yield. The underlying theory is that moisture deficiency in the crop leads to an increase of leaf temperature above air temperature. Thus the term '*stress degree day*' (SDD) has been coined to indicate those days when the crop is suffering from a shortage of water, which will depress yield. The final yield of the crop (Y) can be deemed to be linearly related to the total SDDs accumulated over a given critical period. Thus a general least-squares regression equation (intercept α and β) can be expressed as:

$$Y = \alpha - \beta \left(\sum_{i=b}^{e} \text{SDD}_i \right) \qquad (16.3)$$

where SDD_i is the mid-afternoon (1400 h) value of T_i (leaf temperature) minus T_a (air temperature) on day i, b is the day on which summation begins, and e is the day on which summation ends.

It was first found that the best time over which to sum daily values of the SDD measurements is the period from the first appearance of the awns to the time when the head produces no more dry matter. This period can be

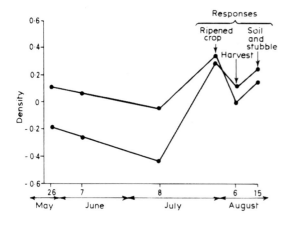

Fig. 16.15 Changes in reflectance of visible light during the growth of spring barley (variety Vada) on soils of the Sherborne Series, Badminton, Gloucestershire. (After Curtis, 1978.)

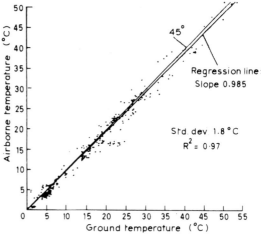

Fig. 16.16 Airborne versus ground-measured temperatures for an entire wheat growing season. (Source: Millard *et al.*, 1979.)

determined by a discontinuity in the albedo data between first head emergence and maturity of the crop. For example, the changes in reflectance of a spring barley crop in England (Fig. 16.15) clearly reflect crop maturity. Albedo measurements during the early stages of crop growth could also provide estimates of the green leaf area index at the time of head appearance. In this way the potential for head growth could be estimated from albedo measurements and then the growth during the crucial period of head growth could be monitored by SDD values obtained by thermal sensing.

There is good reason to believe that temperatures of the canopy of the crop can be measured effectively. For example, a study of a wheat crop at Davis, California, showed that the correlation between temperatures measured on the ground with a PRT-5 Radiation Thermometer (10.5–12.5 µm bandpass) could be correlated closely with airborne measurements at 300 µm altitude using a Texas Instruments R-25 infrared line scanner in the same bandwidth (Fig. 16.16).

Attempts have been made recently to combine the stress degree day concept (moisture factor) with the classical *growing degree day* (GDD) in order to improve crop yield

modelling. The growing degree day can be defined as:

$$GDD = \frac{T_{max} + T_{min}}{2} - T_b \qquad (16.4)$$

where T_{max} and T_{min} are the daily maximum and minimum air temperatures, and T_b is the base temperature (c. 5.5°C) below which physiological activity is inhibited.

Measurement of the climatic factors of GDD plus calculation of daylight minutes in the period between emergence of the crop and the appearance of heads and awns are now used as additional inputs for the estimation of crop yield.

The sensors on the Thematic Mapper and SPOT satellites now available for the study of land use represent a considerable advance on those existing on satellites in the 1970s. The growing experience of remote sensing experts will doubtless refine still further the radiometric characteristics of future satellites. Improved interpretation and classification should be possible as radiometric bands as narrow as 0.04 µm become practicable. The addition of Bands 5, 6 and 7 in the near and

When air photographs are used to study former land use patterns there are two principal lines of evidence. First, there are *vegetation markings*, which mainly occur within cropped land. Where *buried features* exist beneath the surface the soil may be shallow, e.g. over a buried wall. In this case the rooting depth of crops will be restricted and at times of water deficiency (usually late summer) the crops may show yellowing of the leaves or stunted growth. Conversely the buried feature may be in the nature of a ditch. Under these circumstances the soil over the ditch may be more moisture retentive and higher in nutrients. In consequence the crop may be more luxuriant and greener along the ditch line. Such features are sometimes referred to as 'positive' (stronger growth) or 'negative' (weaker growth) crop marks. They are usually best seen when the crop is mature and when the climatic conditions have produced moisture stress in the soil.

The direction of photography is of crucial importance in archaeological survey. Differences of contrast as a result of photography taken in different directions relative to sun azimuth can be striking (Fig. 16.18).

The second type of evidence of prehistoric features consists of *soil markings*. These occur where the soil is bare or has only the thinnest covering of vegetation. The disturbed soil along field boundaries or foundations is different in structure, texture and colour from that of the undisturbed adjoining soil. These differences are often sufficient to induce different reflectances from the surface, which can be detected by remote sensors. In England some of the greatest contrasts have been seen on chalk and limestone soils. Usually soil markings are most clearly detected after a period of exposure to wind and rain rather than immediately after ploughing.

Some archaeological sites can be identified best as a result of transient surface climatic conditions. For example, dark bands of bare earth may reveal where snow or frost has melted first above the filling of buried ditches.

(a)

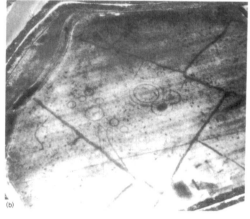

(b)

Fig. 16.18 Crop marks of ring ditches at Lawford (Essex), June 1970: (a) looking into the Sun; (b) with the Sun. Tonal contrasts are entirely due to direction of view since photographs were taken within a minute of each other. (Source: University of Cambridge; Copyright reserved.)

A similar effect is sometimes seen where the lines of buried foundations (or perhaps more probably of trenches where the foundations have been dug out and robbed for later buildings) are picked out by early melting of hoar frost on grass.

The patterns of ancient field systems are often incorporated into present-day field patterns. It is not always easy to distinguish old field boundaries from maps but aerial photography can often provide startling pictorial evidence for such features. For example, Roman field systems often can be observed in great detail. The Roman method of land partition was to allocate blocks of land which were usually in the form of squares (the *centuria quadrata*) of 710 m sides. High-altitude photographs show the grid pattern of centuriation well and in some cases lower altitude observations allow detection of earlier Greek systems to be identified. Similarly the patterns of ancient strip fields can be mapped from remote sensing data (Fig. 16.19).

In this brief discussion of the use of remote sensing for the study of prehistoric land use, attention has been focused on the photographic sensors. However, the use of infrared systems, on both aircraft and satellites, has also provided useful information of an archaeological nature, the latter particularly in broad-area reconnaissance-type studies of areas not yet investigated in detail on the ground. Also there is growing interest in the potential of airborne active sensor systems for such purposes – for example, using synthetic aperture radar designed originally by Jet Propulsion Laboratories in the USA for imaging

Fig. 16.19 A pattern of medieval fields in ridge and furrow near Husbands Bosworth, Leicestershire. Note how the Sun angle casts shadows that serve to emphasize small changes in surface relief. (Source: University of Cambridge Collection; Copyright reserved.)

the surface of Venus. When flown at an altitude of about 7 km over parts of Belize and Guatemala it was able to penetrate the dense jungle canopy to reveal widely distributed patterns of intensive ancient agriculture, leading to a new perspective on classic Maya civilization. All the major satellite-borne imaging systems discussed elsewhere in this book have yielded data of interest to archaeologists, although their interpretation for such purposes is still in its relative infancy. Imagery from the Landsat MSS and TM sensors, SPOT, the Metric Camera Large Format Camera and MOMS, and SIR-A and SIR-B have all begun to support new archaeological research in areas such as the San Juan Basin in New Mexico (archaeological site densities), parts of Alaska (areas of potential archaeological interest in advance of petroleum exploration), Iraq (ancient Mesopotamian canal systems), Egypt (relations between ancient sites and megalithologic units) and the eastern Sahara (buried river systems and associated settlement zones).

Satellite archaeologists have also been much involved with the data from the Space Shuttle's SIR-C radar, with its multiple polarizations and frequencies. Analysed on interactive image processing systems and combined with other types of archaeological data, both traditional and new, such advances should greatly expand our knowledge and understanding of our heritage, not only in respect of rural settlements and land use but also of the built environment, which is the subject of the next chapter.

17.1 GENERAL CONSIDERATIONS

We remarked in the opening chapter that our environment is partly natural and partly of our own making. Although we have already considered some aspects of our imprint on the surface of the Earth, this imprint is most complete in what we may call the 'built environment'. Whether our focus of attention is a single structure in a rural setting, or some great conurbation, it is here that we find the most unnatural features in the world but also some of the most important for us all, because it is here that we live, and from here that we organize our use or manipulation of our surroundings.

Since the scales of unitary features in the built environment are smaller than most of those in the natural environment which satellites have been designed to identify, satellite remote sensing is still, at present, less helpful in this connection than the more traditional remote sensing approaches of air photo-interpretation and photogrammetry. However, we shall see that significant use has been made already of Landsat and SPOT imagery for analysis and assessment of elements and patterns within the fabric of urban and industrial areas. At even more detailed spatial scales, the relatively recent recognition that spectral reflectance curves differ from area to area within urban areas (Fig. 17.1) is opening up some important new uses of multispectral airborne sensor data, both in terms of urban land use mapping, and the evaluation of very local patterns and changes in urban climates, with numerous practical applications. Doubtless, Earth resources, and even environmental, satellites will play an increasing part in monitoring of the built environment in the future, with growing practical significance for such disciplines as urban geography, urban and rural planning, and transport studies.

This chapter is divided into five parts: the first three are concerned with *built environments of increasing degrees of artificiality*, namely in rural, urban and industrial contexts; the fourth is a different kind of overview concerned with *population*, its distribution, and its change; in the fifth, references will be made to the ways in which remote sensing can be applied beneficially in *civil engineering*, i.e. in helping to bring built structures into being, and sustain them safely.

17.2 RURAL STRUCTURES

Since we have already discussed rural land use in Chapter 15, and much of this involves open (i.e. unbuilt) countryside, our present focus must be on settlements themselves, plus some specialized and often quite distinctive aggregates of features, e.g. airports (Fig. 17.2). As in other contexts in the present chapter, remote sensing is a valuable tool for rural survey purposes because it permits relatively cheap, rapid and repetitive mapping even in areas to which access on the ground might be difficult or dangerous. It is convenient to differentiate between a number of contrasting geometrical components of the rural landscape. These are:

1. *point features*, e.g. dwellings, storehouses and animal sheds;
2. *linear features*, e.g. field boundaries, pipes

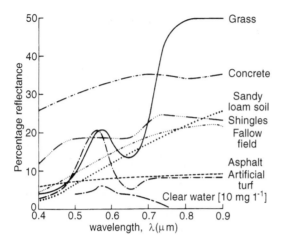

Fig. 17.1 Spectral reflectance curves for selected urban phenomena. The spectra are averages of results discussed in numerous studies. In all cases, data were obtained using a spectroradiometer. (After Jensen, in *American Society of Photogrammetry*, 1983.)

and power lines, roads, railways and canals;

3. *area features*, e.g. settlements (hamlets, villages and small towns), airports, military establishments, large schools, hospitals and penal institutes.

In the case of *point features*, remote sensing surveys may be carried out as aids to topographic mapping. If repeated later, air photography or Earth resources satellite imagery may be analysed to reveal rural change, perhaps for interpretation in economic terms. In this respect a relevant question may be this: is there evidence of changing rural wealth, e.g. through increasing dereliction of homes and out-houses, or through new or improved rural structures?

In the case of *linear features*, such as field boundaries, different sizes or patterns of fields may be found in relation to ethnic and even national differences because of differences in associated cultures and rural practices (Fig. 17.3). Or field patterns may evidence spatial variations in terrain and/or soil characteristics. Significant changes in field boundaries may result from the implementation of land reform policies, especially after a change of government or for more complex agro-economic reasons, such as those responsible

Fig. 17.2 Bristol Airport and surrounding features, 20 April 1975. (Courtesy, County of North Somerset.)

Fig. 17.3 A very clear example of an administrative boundary (the USA–Canadian border), which is revealed in Landsat imagery through associated differences in land use and farming practice, except where topography overrules. To the north of the border lie the grassy central Great Plains of Alberta and Saskatchewan; to the south, the agricultural lands of Montana, where small grains are grown by strip farming methods to reduce soil erosion. In the west the forested mountains of Sweetgrass Arch appear more 'Canadian' than 'American'. (Courtesy, EarthSat Corporation, Bethesda, Maryland.)

for the EC's 'set-aside' policy initiated in the late 1980s and destined to have a growing influence on the use of rural land across much of Western Europe. Remote sensing surveys provide the most convenient means whereby checks can be made on the progress, and effects on the landscape, of any widespread policies of agrarian reform.

The use of remote sensing in respect of existing *communication lines* (roads, railways and canals) is more restricted, for most principal routes have always been mapped in detail and usually with considerable precision. Furthermore, alterations to these networks are usually

controlled by governments or corporations answerable to the general public. Therefore appropriate changes can be incorporated into topographic and route system maps without recourse to additional data sources. However, remote sensing is fulfilling an increasingly vital role in respect of topographic surveys for new road and rail construction projects. Such surveys are widespread, and increasing, especially in developing countries where there is a particularly high premium on a combination of survey accuracy and speed. Even in developed countries, multispectral air photographs and infrared linescan data have long been used by commercial remote sensing survey companies to appraise ground conditions for road and rail planning purposes.

Using remotely sensed data in conjunction with established terrain evaluation procedures it is possible to assess many factors affecting the cost of alternative alignments and types of construction methods. For example, in conjunction with ground truth from selected sites, the evaluation of surface conditions in key localities can be extended along the proposed construction line(s), taking special account of features having practical significance (e.g. slope stability, weathering, erosion, hydrology, etc.). Associated inventories of indigenous materials for road or railroad construction also may be vitally important if the final alignments are to represent the most advantageous and economic routes permitted by the terrain.

The use of aerial photographs and even satellite imagery in supporting *public utility projects* too is more than a future dream. For example, aerial photographs have been used successfully for many years in the development of water distribution services: such photographs provide quick indications of the numbers of premises to be served, the distances involved, and types of land use in the service area. Sewage collection is also efficiently planned thereby. Remote sensing provides perhaps the most practical and economic method whereby gradients can be assessed, the layout of pipelines planned, and pumping stations and force mains appropriately sited. Photogrammetric surveys are always conducted in the USA and other developed countries for the laying of long-distance pipelines and high-voltage transmission lines. In the UK, as elsewhere, the electricity industry regularly checks the condition of major power lines by helicopter-borne infrared linescan. Microwave transmission towers also can be sited best by such means, and their heights selected to meet the necessary requirements for line-of-sight arrangements.

Lastly, in this section, reference must be paid to *area features*, most commonly in the form of small rural community settlements. These may be mapped from air photographs – for example, in exploratory surveys of regions as yet unexplored on the ground (e.g. in tropical rain forest regions such as Amazonia and New Guinea), prior to initial attempts to contact primitive tribes. Similar surveys may be repeated to build up realistic pictures of movements of rural populations still practising shifting cultivation (e.g. in humid tropical forest environments). Radar imagery may be a useful supplement to visible and infrared photography for such purposes, for there is a high incidence of cloud cover in the tropics, and it has been shown that radar images of forested areas may possess internal textural variations that result from a slash-and-burn economy. Indeed, radar is being seen increasingly as a valuable system for delimiting and tracing many artificial features in the rural environment. In rural areas, many frequently recurrent characteristics of elements of the built environment (e.g. flatness, sharpness of corners, and unusual material composition) combine to yield stronger radar echoes from artificial structures than from more natural features of the landscape.

The use of radar can be particularly effective in rural areas when the principles of *polarization* and *cross-polarization* are exploited (see also Chapter 4). We have already seen that all electromagnetic waves are *polarized*; that is to say, once propagated, they continue to move at a given angle measured against a standard plane of reference – unless some outside force or object changes that angle. When radar waves are transmitted horizontally or vertically and are received at the same angle or polarization, they are termed *like-polarized*. If they are received at a different angle, they are said to be *cross-polarized* (section 4.4).

In general, cross-polarized signals produce grainier images than the like-polarized, thus limiting the usefulness of the former. However, under certain circumstances some objects in the rural environment are revealed more clearly in the cross-polarized imagery. For example, detecting and tracing communication nets is performed most easily, completely and accurately using cross-polarized imagery if the net traverses the flight path. On like-polarized imagery the best results are obtained when communication net components are parallel to the direction of flight. Therefore, the most efficient system seems to be one in which both like- and cross-polarized components can be assessed together. The most useful and obvious applications of such a system are in mapping in developing countries. However, it is also possible that, with suitable improvements and refinements to such schemes, urban road systems may be investigated with special reference to their surface materials – an application of considerable practical value to many developed countries also. It is to urban areas that we may now turn our attention in greater detail.

17.3 URBAN AREAS

Whilst urban centres afford the greatest concentrations of facilities for living and working, they also present a wide range of attendant problems. These most notably include *internal* ('intra-urban') problems concerned with the provision of acceptable residences, transportation networks and places of employment, in conditions of suitable freshness and cleanliness, and *external* ('inter-urban') problems concerned with the maintenance of acceptable relationships between one city and another, and between the cities as a group and the intervening countryside. It has been suggested that many of the contemporary problems of cities may be traced to:

1. the reasons for city foundation and growth, and the different histories of city evolution;
2. the ongoing, competitive and conflicting processes of arrangement and rearrangement of land use and functional areas within cities;
3. the scales, complexities, and patterns of concentrations of intra-urban activities.

Today there is widespread interest in the development of 'urban information systems' for the collation, interrelation and practical utilization of urban data from a wide variety of different sources. Urban geographers and planners already recognize remote sensing data as a vital part of the total information pool. Here remote sensing data contribute to two classes of information in particular:

1. *Static phenomena.* These include such things as city size; the number, pattern and capacity of roads; building sizes and types; and the characteristics of types of neighbourhoods (e.g. industrial, residential, commercial).
2. *Dynamic phenomena.* These include variables that cannot be observed directly, either because they change so rapidly, or because they are not physically visible, e.g. population statistics, (aggregated) traffic flow patterns data, and socio-economic conditions.

We will now illustrate and exemplify the value of remote sensing as a source of valuable data for urban information systems by reference to a number of different aspects of the urban environment.

17.3.1 URBAN SPATIAL STRUCTURE AND SETTING

Here the interest focuses on the way in which cities are laid out, and how they are related both to each other and to any smaller intervening communities. Cities can be analysed in terms of hierarchies, which take account of the rank orders of cities, and the distances between them on the ground. Small-scale satellite imagery can be used to establish general hierarchical relationships between cities and towns. For example, night-time visible imagery from the low-light intensifier camera system on some of the early Defense Meteorological Satellite Program (DMSP) weather satellites have been interpreted in this way over areas of subcontinental scale, and are now systems, chemical devices and multi-spectral sensors, capable of providing data with resolutions of between 1 and 10 m. From such data, maps of the physical linkages and terminal facilities in a network could be compiled. Pattern analyses by computer processes could be carried out once the information had been transformed into matrix and vector form.

2. *Flow phenomena and associated problems.* Here infrared and panchromatic photography and radar could be used, giving an optimal resolution of 0.5 m. Matters commensurate with investigation would include origin–destination patterns and daily 'tidal' flows of traffic as a whole. These have useful applications in road design and traffic control.

3. *Transport/land use interrelationships.* These necessitate infrared, colour and panchromatic photography, and chemical sensing devices particularly sensitive to phosphorus and nitrogen. The optimal resolution would be 1 m. Matters such as the relationships between land use intensity and distance between road and rail links, land use capability, and the positions of areas within their larger economic regions should be amenable to study.

The list indicates something of the potential of the satellite in transportation studies as well as some of the hopes and aspirations of would-be data users, which are important stimuli for further advances in remote sensing technology, science and operations.

Clearly the ranges and resolutions of remotely sensed data generally available at present are not yet adequate for the achievement of such aims – but this situation will change quickly in the next few years (see Chapter 19).

17.4 INDUSTRIAL COMPLEXES

In view of the great economic, social and political significance of manufacturing and extractive industries, and the dominant role they play in many urban areas, we may consider these in more detail, both for their own sakes and in order to exemplify in greater detail methods of application of remote sensing data to the built environment. Three aspects of the industrial components of urban areas may be elucidated most usefully by aircraft or satellite imagery:

1. the present location of industry, and different types of industrial land;
2. heat loss and the spread of atmospheric and water pollutants away from an industrial complex;
3. opportunities that exist for the establishment of new industries, e.g. for the redevelopment of land under existing industrial use.

Historically, the analysis of industrial activity from remote sensing imagery can be traced back to World War II, when air photography was used for the identification of tactical military targets and the selection of military objectives for aerial attack. More recently, various types of airborne and spaceborne sensor data have been used to support the offensive actions of the Coalition Forces in the Gulf War of 1990–1991. Today, the analysis of remotely sensed imagery of industrial areas has many peaceful uses also, not only in mapping industrial areas and their changes through time, but also in associated activities, such as updating inventories of stockpiled materials.

Aerial photography is still the most commonly used remote sensing imagery for industrial analysis: usually high-resolution data are essential for the adequate interpretation and assessment of features of the industrial landscape. Aerial photography of industrial areas is frequently recorded at scales larger than 1:10 000, even in some cases larger than 1:5000. Colour film is sometimes used in preference to panchromatic film on account of the higher definition it provides. Often photo-interpretation keys for use in industrial areas are very detailed and technical; their compilation is a highly skilled business, requiring intimate knowledge of industrial structures and processes.

Stockpile inventories may be made more accurately and expeditiously from air photographs than from ground where raw materials or finished goods, such as coal, ores and other minerals, pulpwood and lumber are accumulated in the open air. Neatly stacked materials can be assessed very accurately, albeit only from air photographs at scales of 1:1000 or better. Where stockpiles are of less regular shapes more variance is to be expected in the periodic estimates. However, some experts claim to be accurate within 2%, and assert that the lowest-cost ground estimates have an average accuracy of only 15%. The same experts emphasize that air survey costs can be substantially below those of surveys on the ground. Since both methods follow essentially the same set of procedures, they can be used together or interchangeably. It is commonplace to:

1. map the main stockpile with closely spaced contours (often only 1 or 2 m apart);
2. map the tops of the piles (whose areal dimensions are smaller) in greater detail;
3. determine stockpile volumes by planimetering areas between successive contours and multiplying by the depth of material;
4. convert the cubic unit volumes into weights, using tables for specific minerals;
5. adjust the initial estimates of weights using compaction factors to allow for settling of material in the pile(s).

Similar kinds of measurements can be made in support of extractive industries, e.g. of the amounts of material removed from an opencast mine through selected periods of time. The resulting data can be of great value to mining companies and, potentially, to governments seeking an accurate, low-cost means of monitoring mineral industries for taxation on a weight basis.

Prompted by the major fluctuations in global oil prices that have taken place in recent years, much interest has been shown in the use of infrared linescan imagery for the mapping and assessment of *heat loss from buildings* (Plate 13, colour section). Poorly insulated buildings are often quite clearly evident even in the raw infrared image products. By thermographic mapping, values can be placed upon the rates of heat loss through every structure covered by a survey. Many commercial remote sensing firms today provide airborne heat-loss survey services. Indeed, these are widely used not only by industrialists, but also by bodies such as central and local government, housing authorities, hospital management boards, etc.

When we broaden the discussion to embrace the whole field of environmental *pollution monitoring*, it emerges that a wide range of remote sensing techniques have now been tested in the search for efficient methods to measure the concentration and spread of pollutants. Examples include:

1. *Visible waveband photography*. This has been used for such studies as the behaviour of chimney plumes and the appearance and intensification of urban, grassland and forest fires.
2. *Infrared imagery*. This has been used as the basis for studies as diverse as the monitoring of heat loss from buildings, forest fire detection, stubble burning, and the delineation of warm effluents in rivers, lakes or the sea.
3. *Microwave techniques*. These are increasingly being used to map oil spills in coastal and deeper waters.
4. *Multispectral processing techniques*. These can be used to assess water quality through the different levels of light penetration in selected regions of the thermal emission spectrum.

In general, it has been suggested that remote sensing of such aspects of environmental quality in and around urban centres and elsewhere may have two significant contributions to make: first, through the provision of a more complete spatial inventory of areas of

382   *The built environment*

estimate population densities and totals for small areas within larger urban centres, using multispectral air photographs and ground truth from sample areas. One such model was applied to Washington, DC, in 1970. It was formulated as:

$$Y = f(x_1, x_2, \ldots, x_n) \qquad (17.1)$$

where Y represents housing unit counts or population (1970 tract statistics) and $x_1 - x_n$ are imagery derived variables, such as the number of single-family structures, the number of multiple-family structures, distance from the central business district, etc. Relationships between x and Y were developed by multiple regression analysis.

In practice, residential land use was identified on 1:50 000 scale aerial photographs, and the residential land area was computed. A block-by-block count was then conducted to establish the number of dwelling units per structure according to a four-category classification (single-family housing units, 2–5, 6–14, and more than 15 housing units). Multiple

regression coefficients of 0.65 and 0.54 were obtained for central city and suburban tracts, respectively. These were encouraging, but pointed to the need for a number of improvements before such a scheme became operational. These included the provision of remote sensing imagery of a type in which tree cover poses less of a problem for the location and classification of housing units, plus a greater range of ground truth data for the formulation of the model and the checking of its output.

We may conclude that remote sensing, even at a coarse scale, can clearly provide valuable information concerning population levels and changes in most areas of the world. The application of aerial photography to dwelling-unit and population estimation would appear to have a particular potential in developing countries where rates of urban and population growth are especially great. Demographic data in such countries are frequently less than adequate for planning purposes, since national censuses are difficult to undertake and often yield inaccurate results.

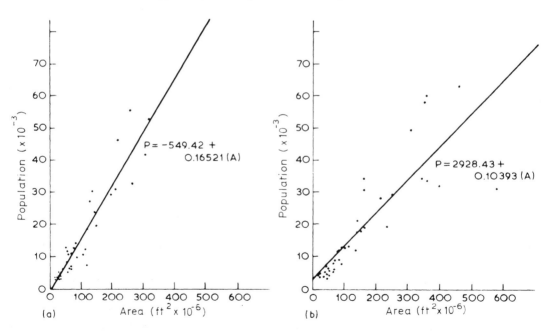

Fig. 17.6 Observed relationships between population and area of towns in the Tennessee River Valley, in (a) 1953 and (b) 1963. (Source: Holz *et al.*, 1969.)

Table 17.2 A comparison of housing areas in Austin, Texas, based on selected environmental criteria amenable to identification on remote sensing imagery (Source: Davis *et al.*, 1973)

Criteria	Low-income areas	Middle-income areas
House size (ft^2)	380–1220 (average 731)	110–1560 (average 1305)
Placement of house and lot (distance from street in feet)	12–42 (average 27)	34–45 (average 40)
Potential landholding per housing unit (ft^2)	5670	7500
Building density (%)	14.2	17.7
Average lot size and frontage (ft^2)	4337	7376
	Narrow frontage	Wide frontage
Image, pattern, and texture	Lack of uniformity	Uniform
	Irregular	Regular
Houses with driveway (%)	8.3	97.0
Houses with garage (%)	3.0	97.0
Number of visible automobiles per house	0.20	0.76
Unpaved street (%)	65	0
Street width (ft)	12–24 (average 18)	24–31 (average 28)
Quality of curbing	Generally lacking	Intact
Quality of vegetation	Not as vigorous	Cultivated, vigorous
Housekeeping	Presence of debris	Lack of debris
House orientation to street	Short side	Long side
City block pattern	Irregular, dead end twisting streets	Regular
Vacant lots per city block	0.6	0
Proximity to manufacturing and retail activity	Close to manufacturing Remote from shopping manufacturing	Remote from manufacturing Close to retail outlets

A partly related question of considerable significance to government and municipal authorities is that of *housing quality*. For example, concerted attacks on the problems of urban poverty neighbourhoods cannot be planned – still less carried out – until such areas have been defined, and located on the ground. Conventional methods of amassing data on housing quality distributions are extremely time consuming. Once again remote sensing techniques may provide new and better data than surveys on the ground, while providing such data more quickly and frequently.

Table 17.2 presents a classic comparison of housing areas in Austin, Texas, based on selected environmental criteria. Most of these features may be assessed from remote sensing imagery under normal viewing conditions. We may appraise housing quality, study certain socio-economic aspects of a neighbourhood, and estimate the level and distribution of family income on the basis of such a range of urban characteristics.

17.5.2 CHANGES IN THE BUILT ENVIRONMENT

We have seen how remote sensing systems may be exploited for both rural and urban land use detection; without doubt, remote sensing data are of greatest value when analysed for *landscape change*. It should be noted, too, that because such data are unselective (given a particular image type with its attendant spectral and spatial resolutions), the remote sensing view of a built environment reveals it as a *single system, organically related to the open land within and around it*. Since cities are the most powerful and vigorous built environments we may conclude this section with reference to a range of methods and projects designed to monitor their growth and changing impacts on their own environs. Although remote sensing data can be analysed manually to provide rough estimates of urban growth and associated landscape change, there is again a trend towards the development of increasingly sophisticated automatic interpretation procedures for such purposes.

One early and ambitious project to evaluate the usefulness of photography from high-altitude aircraft and Earth-orbiting satellites for urban land use change detection was the 'Census Cities Project'. This was organized by NASA in conjunction with the Geographic Applications Program (GAP) of the US Geological Service (USGS). Originally 26 cities were named as test cases, and the US Air Force Weather Service and NASA's Manned Spacecraft Center acquired multispectral, high-altitude photography from 20 of them. The year chosen for this coincided with, the time of the 10-year US Census. The basic remote sensing data were colour infrared photographs. This film type is especially valuable for use over towns on account of its high haze-penetration capability. A further useful characteristic of this type of film is that it distinguishes clearly between vegetation (reddish in colour) and cultural features (which have a blue appearance on the film). This makes it especially useful where cities have a strong urban structure/woodland mix. The ultimate goal of the Census Cities Project was the production of an Atlas of Urban and Regional Change, accommodating many types of data presentation, including photomosaics, conventional maps, computer-printed maps, tabulated data and text. In a pilot study for the city of Boston, a 24-category land use classification was used. The minimum cell size was about 10 acres (4 ha). The land use data were computer processed to make them compatible with the census data. Consequently they can be retrieved either singly, or in selected combinations, by the census tracts.

On a smaller scale, Landsat imagery has been processed to give land use maps of urban changes at scales from 1:250 000 upwards. One such programme brought together the Planning Intelligence Department of the UK Department of the Environment (DoE), and the Image Analysis Group at the UK Atomic Energy Authority (UKAEA) at Harwell, Berks. For planning purposes, land use information is needed especially around urban areas, but, at present, there is still no single source of comprehensive data in the UK which can provide up-to-date information of this kind on a national scale. Consequently studies have been undertaken to investigate the potential of Landsat as a suitable information source. Using imagery obtained from the European receiving station at Fucino, Italy, a NASA software package has been further developed for quick and efficient geometrical rectification. Each Landsat scene is computer processed using a multicategory supervised classification in multidimensional measurement space. To verify the suitability of Landsat MSS data for monitoring urban growth, four test areas (each about 35 km^2) covering a wide range of conditions were examined in detail. The accuracy of computer classification of the Landsat data was checked against ground truth obtained from aerial photography and Ordnance Survey maps. Taking Northampton, an important English Midlands town (population 130 000) as an example, over 90% of new development that took place from 1969 to 1975 was accurately identified (Fig. 17.7).

More recently still it has been shown that urban areas can be classified and mapped with even greater success using the higher spatial resolution data from the Landsat TM and SPOT multispectral or panchromatic systems. However, it is clear that greater accuracies would be achieved if ground survey classifications were better suited to the characteristics of the satellite data; traditional land use classifications cannot always be applied very successfully to either supervised or unsupervised classifications of satellite imagery.

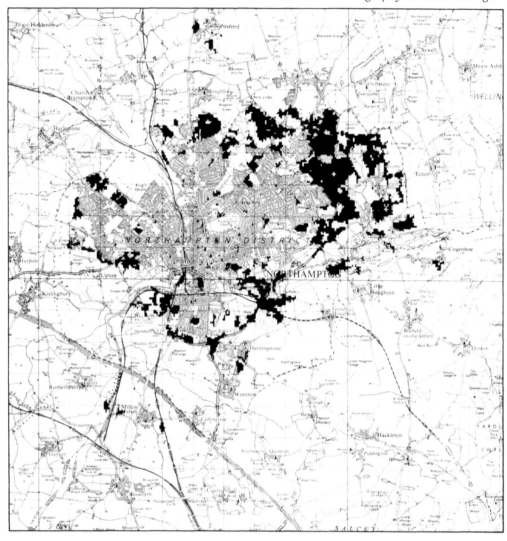

Fig. 17.7 Urban growth (shaded black) around the English Midlands town of Northampton, beyond the 1969 developed area boundary, as determined from Landsat (1975) imagery. (Courtesy, UK Atomic Energy Authority and the Department of the Environment.)

However, information obtained from satellite studies can be compared and integrated easily with digital data of other kinds (e.g. from population censuses) for many central and local governmental planning applications, and is increasingly being used in support of commercial planning and development, e.g. in relation to the selection of sites for new supermarkets, filling stations, fast-food outlets and many more.

Turning finally to the change-detection potentialities of remote sensing data, studies of general user requirements for land use data in the urban environment confirm that, *for different types of user, different scales of analyses are necessary*, and, therefore, different types of remote sensing data must be used. For example, if urban land use information is required only on a grid square or block basis, then a relatively low-resolution system will provide adequate information and be advantageous in eliminating the finer details of urban morphology, which may then be regarded as a form of picture 'noise'. Aerial photography or satellite imagery, e.g. from Landsat TM or SPOT, supporting maps prepared at a scale as small as 1:100 000, may then suffice. On the other hand, if details of land use at the plot or parcel level are sought, then much larger-scale data are necessary, supplemented by additional types of information. Some of the most exciting new work in this field is permitting the generation of 3D city models, which can be investigated via 'virtual reality' (Fig. 17.8).

Governments are now beginning to set up formal procedures for the operational inclusion of remote sensing data into urban and regional land use change-detection schemes. The general form such schemes are taking is shown by Fig. 17.9. Remote sensing is vital to it, since this provides the 'status report' information involved in the 'update' section of the scheme in operational use. Clearly, frequent snapshot data from high-altitude platforms should increasing the dynamic element in studies of urban land use and morphology, with resulting benefits to the whole population through more economic and rational control, and/or development of the complex urban/rural systems of which the modern nation is composed.

In the same context, mention should be made of the role of remote sensing in the emerging sphere of interest in urban systems as *ecological entities* – with relevance to the need to improve cities as the homes of both human and other living populations (*vide* the new *Journal of Urban Ecosystems*). Remote sensing is increasingly able to provide detailed information on many changing biophysical properties of urban areas (e.g. ratios of impervious, vegetated, water and soil surfaces) to support improvements in the modelling of affected processes (e.g. storm run-off, evapotranspiration, thermal gains and losses, pollution chemistry, water uptake by vegetation) and predict and manage their impacts on urban populations. Figure 17.10 schematizes some of the interrelations in this very contemporary field.

17.6 REMOTE SENSING FOR CIVIL ENGINEERING

Although we have noted some civil engineering applications of remote sensing in passing in the earlier sections of this chapter, this activity is so important in the planning and development of the built environment that it deserves specific mention on its own. For at least 50 years the interpretation of aerial photographs has been an indispensable technique for accomplishing a wide variety of engineering projects. In recent years newer remote sensing systems, such as infrared linescans, radars and satellites, have become increasingly important sources of information of value to the civil engineer, not only through the *mapping* information they provide for relatively static features of land areas and coastal zones, but also through their *monitoring* capabilities with respect to weather, ice and snow, water currents, lakes and rivers, and other relatively dynamic phenomena.

Although each engineering project is more or less unique, the chief functions and basic methods of image utilization are similar in all projects. These may be summarized as follows:

1. Preliminary planning:
   - (a) listing of the engineering survey purpose(s) and the prime factors which must therefore be evaluated;
   - (b) reviewing the literature and the existing remotely sensed data;
   - (c) planning for the new image data/results acquisition.

2. Data collection:
   - (a) performing preliminary field investigations to determine reflectance, emittance and other key properties of natural materials and critical terrain features, to help to select types of data required, and the tools and methods of analyses;

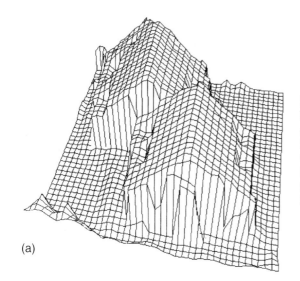

(a)

Fig. 17.8 Towards 3D city models based on air-borne laser data and 2D digital GIS data (digital cadastral map): (a) sample digital stereo model (DSM) of typical houses; (b) 3D visualization of virtual city model, produced by mapping terrestrial images against DSM building faces, and adding artificial trees. (Courtesy, Haala and Brenner, Stuttgart University, 1998.)

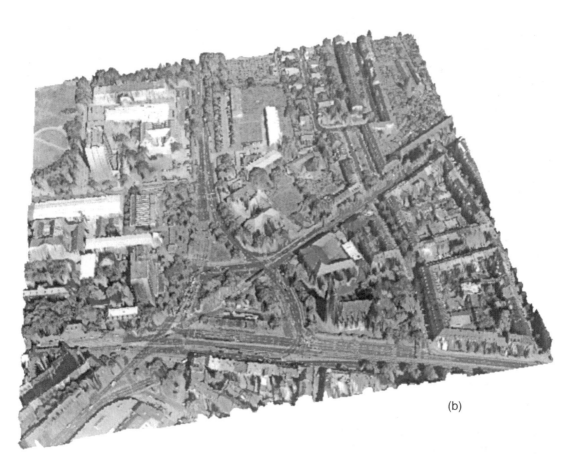

(b)

(b) selecting best weather conditions for data collection;

(c) obtaining imagery as planned;

(d) collecting collateral data simultaneous with aircraft/satellite overpasses.

3. Data analysis:

(a) merging and display the remote sensing data for the whole area to provide an overview of the study region;

(b) reviewing the collateral data to help identify pertinent features;

(c) developing a regional concept of the study area based on the above, and a growing understanding of the features of significance;

(d) performing detailed analysis of the remote sensing data, e.g. With respect to topography, drainage, erosion, vegetation, and cultural patterns;

(e) developing a classification relevant to the project and define the basic units to be mapped and/or monitored;

(f) delineating the basic units classified in the data set(s).

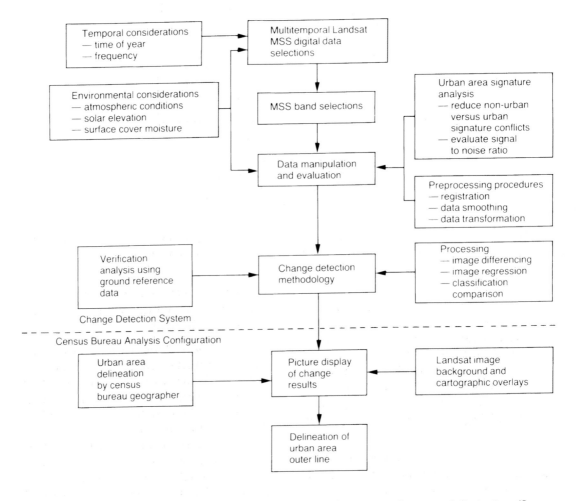

Fig. 17.9 Flow diagram of an urban change detection system for use in urban area delimitation. (Source: Toll, 1980.)

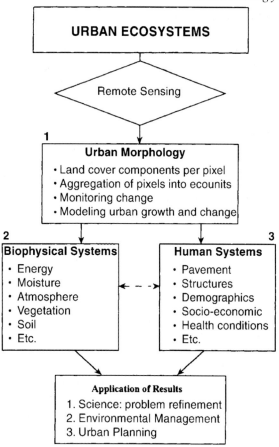

Fig. 17.10 A three-part role for remote sensing in 'urban ecosystem studies', the results of which serve the purposes of science, environmental management, and urban planning. (Source: Ridd, 1995.)

4. Field verification:
   (a) selecting representative sampling locations, and routes of travel;
   (b) performing field verification, bringing remote sensing data and/or products into the field for detailed correlations;
   (c) collecting samples to help to refine knowledge of basic surface properties;
   (d) completing laboratory testing of field samples.

5. Final analysis and presentation of results:
   (a) use of field verification data to correct and complete the original analysis and prepare the final product;
   (b) preparation of final report.

The most frequent and important types of civil engineering studies which have used remote sensing include:

1. *Terrain analyses.* Almost every major construction project is controlled by the natural terrain: it is rare that some adjustment of terrain is not required, e.g. through cutting, infilling, banking, draining or reclaiming. The cost of moving the natural materials is usually the largest single cost in any engineering construction project. The main value of remote sensing imagery in this respect is that it provides a wealth of detail in the form of a spatially continuous model of the landscape, often able to reveal

features not evident to the naked eye.

2. *Mapping 'landform engineering soil characteristics'*, especially soil texture, drainage conditions, and slope categories. For many engineering studies air-photos may even be used as the mapping bases in preference to existing topographic or soils maps.

3. *Preparation of construction material inventories*, especially inventories of granular materials (either naturally occurring unconsolidated deposits, or consolidated deposits suitable for crushing). Air photos have been used for many years to identify and assess such deposits; more recently both colour infrared and Landsat MSS data have been used for the same purposes.

4. *Preparation of regional materials inventories*. Regional inventories of construction materials have been prepared from Landsat data to distinguish between major type categories, e.g. organic, alluvial, littoral and glacial deposits, gravel pits, Precambrian rock, Palaeozoic rock and lineaments at scales of 1:100 000, supplemented by colour infrared images from airborne surveys for key areas at scales of down to about 1:25 000.

5. *Site investigations*, whether for drains and reservoirs, bridge and pipeline crossings of rivers, airstrips and airports, staging and docking sites, industrial, town or recreational sites, etc. Within the last two decades interpretation of conventional (panchromatic) air photographs has been increasingly complemented by recourse to colour and colour infrared air-photos, airborne infrared linescan and radar imagery, plus Landsat and SPOT data, through a hierarchy of scales (Table 17.3). In future the use of imaging spectrometers may be expected to become important too.

6. *Water resources engineering*. Perhaps more than any other area of engineering this demands information of an environmental monitoring nature as well as an Earth resources mapping nature. Many types of remotely sensed data are directly applicable, e.g. through watershed analysis, surface water mapping, snow quantification, flood damage quantification and even groundwater assessment. Increasingly efforts are being made to bring these and *in situ* and collateral data sets together in a geo-based information system. The same is true in river, lake and coastal water environments, although here data from other specialized sensors, including the Nimbus-7 CZCS, the SeaWiFS ocean colour sensor and ERS SAR, Altimeter and even Scatterometer types of instruments, become important.

Table 17.3 Types and scales of remote sensors for different stages of site investigation

Stage of investigation	Type of sensor	Range of scale
Large area evaluation	Air-photo mosaics and imagery, two-dimensional (Landsat) and three-dimensional (SPOT) viewing Earth resources satellite	1:100 000 to 1:80 000
Selection and evaluation of alternative sites	High-altitude black and white panchromatic aerial photography (stereoscopic viewing)	1:80 000 to 1:40 000
Intermediate stage terrain examination and evaluation of selected site	Black and white panchromatic and/or colour infrared aerial photography (stereoscopic viewing)	1:40 000 to 1:20 000
Detailed terrain classification and mapping followed by site location	Conventional black and white panchromatic, colour, or colour infrared aerial photography (stereoscopic viewing), plus radar surveys	1:20 000 to 1:10 000

7. *Transportation facility planning and construction*, involving such matters as highway planning, route location, traffic surveys, railroad construction and maintenance, slope stabilization, etc. In these contexts satellite imagery may predominate in developing countries, but airborne data in developed countries.

8. *Litigation and claims.* Areas in which remote sensing imagery (mainly from aircraft) have been used extensively for such purposes have included condemnation cases to help to determine the value of land and appurtenances; damage suits, e.g. as a result of a highway accident; and environmental impact hearings, e.g. in relation to the construction of some new building(s) or complex of buildings and their anticipated effects on topography and landscape. In such contexts remote sensing data are important because they are permanent and unbiased records of all natural and cultural elements, as revealed by their intrinsic characteristics.

The only possible conclusion to this chapter is that remote sensing applications relating to the evaluation, management and planning of the built environment are amongst the *most directly and obviously cost-effective of any of the applications covered by this book.* Remote sensing is capable of delivering relevant information that is often more abundant and timely than that which is otherwise available from all other sources.

18.1 DEFINITIONS

We live in an age of growing sensitivity to environmental traumas. As population densities continue to increase, living standards and their expectations rise, and the value of human property chronically inflates, so the impacts of abnormal and extreme events in the environment become ever more serious and costly. Today environmental traumas are both numerous and very varied, as Table 18.1 exemplifies. Many of these events are natural; more and more, though, are *anthropogenic*, i.e. induced by humans, for example by acts of war and terrorism. Some anthropogenic traumas are caused unconsciously or indirectly: for example, landslides resulting from slope destabilization by the excavation of road or rail cuttings, undesirable soil changes resulting from vegetation clearance and subsequent agricultural activities, and climate changes related to increased emissions of greenhouse gases into the atmosphere and damage to the ozone layer induced by chlorofluorocarbons (CFCs), as discussed in Chapter 1. However, natural traumas probably still constitute the biggest threats to us and our economy. It is on these that we will concentrate most.

First we must define our terms.

1. A *hazard* is a condition (natural or anthropogenic) of the environment which can exert an adverse influence on human life, property or activity. Since we have to a large extent geared ourselves to the statistically normal state of our environment, the most significant hazards usually accompany environmental states or phenomena which are liable to high degrees of variance about the mean.

2. A *disaster* is a serious, damaging effect on human life, property or activity which results from the impact of a hazard that has exceeded its critical level(s).

Table 18.1 provides a classification of common causes of environmental disasters. It was estimated in 1978 that the cost to the global economy of natural disasters was at least US$ 40 billion; today, the figure must be about an order of magnitude larger. Table 18.2 lists those known to have resulted in economic losses of more than 1% to national economies between 1960 and 1987. It should be noted that Table 18.2 refers only to 'sudden impact' disasters: many others of similar or greater severity

Table 18.1 Common causes of environmental disasters (After Galli de Paratesi, in Barrett *et al.*, 1991)

Geology	Meteorology
Earthquake	Precipitations
Volcanic eruption	Severe local storm
Landslide, mudflow	Tropical storm
Tsunami	Storm surge
Erosion	Severe winter freeze
	Frost and freeze
Hydrology	Vegetation
Flood	Fire
Flash-flood	Plant disease
Drought	Drought
Avalanche	Insects
Snow	
Oceanography	
Coastal flooding	
Marine pollution	
Relative sea-level rise	

occurred during the same period, but these were 'creeping' disasters, i.e. slowly developing and/or chronic situations (e.g. drought-induced famine) whose costs (e.g. through increased food imports, loss of productive manpower, etc.) are much more difficult to estimate. The cost of natural disasters to the world economy is on an ever-rising trend. Nearly one-half of the astonishing costs involve expenditure on *disaster prevention*, including the monitoring of potentially serious environmental hazards. Clearly, whilst hazard monitoring and disaster assessment procedures will play an increasingly vital role on the world stage in the twenty-first century, every effort must be made to ensure that they are carried out not only as efficiently, but also as economically, as possible. Remote sensing is highly promising in both respects.

18.2 DISASTER ASSESSMENT

Today most large nations, and many international agencies and organizations, are specially equipped to respond to disaster situations. The United Nations itself has a practical interest in such matters, spearheaded by UNDRO (the United Nations Disaster Relief Organization), with headquarters in Geneva, Switzerland, supported by several other UN Agencies, plus numerous international and national relief bodies. Once a disaster has occurred, the priorities

Table 18.2 List of major natural disasters, which caused economic losses of more than 1% of GNP, 1960–1987 (Source: Galli de Paratesi, in Barrett *et al.*, 1991)

Country	Event	Date	Deaths	Loss (10^6\$)	GNP (10^9\$)
Morocco	Earthquake	2–60	13 100	120	12
Chile	Earthquake	5–60	3 000	800	17
Yugoslavia	Earthquake	7–63	1 070	600	45
Philippines	Typhoon	11–64	58	600	32
Italy	Earthquake	5–76	978	3 600	352
Peru	Earthquake	5–70	67 000	500	17
Nicaragua	Earthquake	12–72	5 000	800	3
Honduras	Hurricane	9–74	8 000	540	3
Guatemala	Earthquake	2–76	22 778	1 100	9
Italy	Earthquake	5–76	978	3 600	352
China	Earthquake	7–76	242 000	5 600	280
Romania	Earthquake	3–77	1 581	800	51
Yugoslavia	Earthquake	4–79	131	2 700	45
Caribbean/USA	Hurricane	8–79	1 400	2 000	?
Algeria	Earthquake	10–80	2 590	3 000	47
Italy	Earthquake	11–80	3 114	10 000	352
Greece	Earthquake	2–81	25	920	33
Yemen	Earthquake	12–82	3 000	90	4
Peru/Ecuador	Floods	4–83	500	700	27
Fiji	Cyclone	3–83	7	85	1
Colombia	Earthquake	3–83	250	380	35
Chile	Earthquake	3–85	200	1 200	17
Bangladesh	Cyclone	5–85	11 000	?	?
Mexico	Earthquake	9–85	10 000	4 000	136
Colombia	Volcano	11–85	23 000	230	35
El Salvador	Earthquake	10–86	1 000	1 500	4
Iran	Floods	12–86	424	1 560	90
Vanuatu	Typhoon	2–87	50	200	0.1
Ecuador	Earthquake	3–87	1 000	700	10
Bangladesh	Floods	9–87	1 600	1 300	12

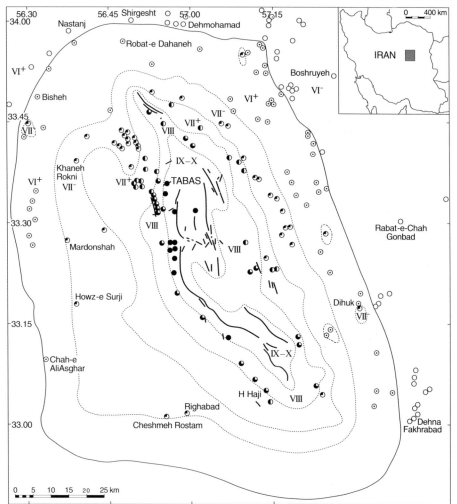

Surface rupture with 1968 earthquake; Isoseismal on Modified Mercally Intensity scale:
●80–100% destruction ◐50–80% ◑35–50% ◒20–35% ⊙5–20% ○5% destruction

Fig. 18.1 Isoseismal map and aircraft-assessed damage distribution in the epicentral region of Tabas-e-Golshan (Iran) earthquake of 16 September 1978. (Source: Berberian, 1978.)

for action include *assessment* of the damage caused, *relief* for the affected population, and *planning* of measures necessary to return life to normal in the disaster area. The use of remote sensing in such connections is already well established. Aircraft surveillance of stricken areas is often the most rapid and convenient way of building up a realistic picture of the scale and the intensity of a disaster situation. Air survey and photogrammetry is often useful

in the formulation of plans to facilitate safe reoccupation and reuse of the affected region. Figure 18.1 illustrates the value of aircraft remote sensing in disaster situations. Often this is still the only way of obtaining a quick over-view of the troubled zone, for surface commu-nications may be severely affected, though as satellite systems improve in respect of both the range and timeliness of their products, these will become increasingly vital for the

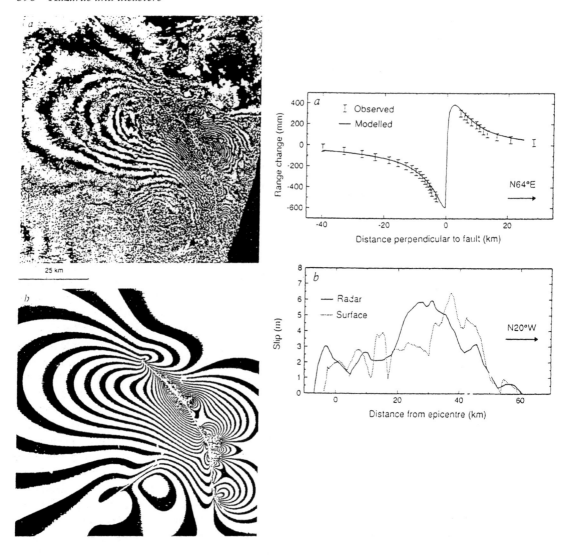

Fig. 18.2 ERS-1 SAR interferogram (top) and model (below) of the Landers Earthquake in the USA, as observed by the satellite (top left), and modelled from those observations (bottom left). Surface observations coincided well with the satellite-derived pattern (see graphs on right). (Courtesy, ESA.)

monitoring and modelling of disaster events, e.g. through the use of satellite-borne radar in respect of some types of earthquakes (Fig. 18.2).

As the range and ability of sensors for remote sensing has improved, so the range of disasters that can be evaluated by airborne systems has expanded too. By way of example, reference can be made to the special development programme undertaken by the Institute of Machinery, Faculty of Engineering, University of Catania, Sicily, in relation to disastrous and potentially disastrous events in southern Europe and the Mediterranean Sea.

The *platforms* they have developed or exploited for such uses include:

1. fixed platforms (ground-based masts and towers);

2. fixed-wing aircraft, e.g. twin-engine Piaggio P.166-DL3;
3. helicopters, e.g. Agusta A-109A, Agusta Bell AB-204B, AB-212 and AB-412 and Agusta Sikorski SH-3D;
4. naval patrol craft.
5. Earth-orbiting satellites, including Landsat, SPOT and NOAA.

The *sensors* they have used include:

1. photographic, e.g. Aerial Camera System (ACS) and Vinten 618;
2. ultraviolet and infrared scanners, e.g. Daedalus AA 2000 Bispectral scanner;
3. visible and near-infrared scanners, e.g. Landsat MSS and TM and Daedalus DS1268 Multispectral scanner;
4. thermal infrared scanners, e.g. FLIR Systems 2500 A/F;
5. lasers;
6. radar;
7. satellite-borne visible, infrared, multispectral and passive microwave radiometers.

Several of these sensors have been combined with an airborne platform in the University of Catania's helicopter-based 'Pelican System'. This system includes provisions for multispectral photography (including simultaneous black and white infrared, colour infrared, colour and ultraviolet imaging of the same scene), thermal infrared imaging and digital image processing.

The system has been designed to monitor and protect the marine environment, which has been the primary concern of the Catania laboratory. Specific applications of the package have included:

1. oil spills from ship collisions;
2. illegal discharges from ships;
3. oil slicks, thermal and other types of marine pollution;
4. urban and industrial discharges into rivers and coastal waters.

A recent cost analysis of major oil spills, reported by the International Tanker Owners Pollution Federation, indicates that costs of clean-up operations after tanker accidents range from about US$5000 to US$30 000 per metric ton of oil spilled. Additionally claims for damage to the environment following major disasters of this kind may reach many tens of millions of dollars. It has been estimated that a regular use of multisensor remote sensing techniques during oil-spill response phases should reduce clean-up costs by at least 15–20% – whilst reducing general damage to flora, fauna and the environment as a whole by a very significant extent.

Although there is increasing interest in the potential of satellite surveillance for disaster assessment, that potential is sometimes difficult to realize. For example, present civilian weather satellites lack both the high spatial resolution capability, which is often required for such purposes, and the all-weather (microwave) capability, which would enable them to penetrate cloud cover and reveal the situation on the ground irrespective of the state of the sky. Further, present Earth resources satellite types lack the high temporal frequency of imaging which is generally necessary for the evaluation of disasters: typically, these strike with great suddenness and coincide relatively infrequently with the availability of suitable Earth resources satellite data for the area(s) in question. Perhaps the best hope for the serious use of satellites in disaster assessment in the foreseeable future involves the notion of *steerable high-resolution sensor systems on gestationary satellites*, if these could be pointed, on demand, towards any area of key significance. The polar-orbiting French SPOT satellites are equipped with a steerable sensor (Chapter 5), but operate in a low non-geostationary orbit, and therefore lack the very frequent revisit capability that disaster assessment really demands.

18.3 HAZARD MONITORING

It is clear, then, that satellites have a much more immediate and extensive part to play in hazard monitoring – and the international community appreciates the overwhelming logic of improving hazard monitoring in order to reduce the impact of hazards when they 'go critical' and a disaster threatens. At the same time, of course, it is clear that, whereas aircraft are better than satellites for disaster assessment, they are much less efficient and cost-effective than satellites for hazard monitoring. Aircraft operations are notoriously expensive, ranging at present from some US$500–10 000 per hour in the air, and are totally inappropriate for assessment of widespread conditions, e.g. snowfall. For these conditions in particular, satellite mapping and monitoring is highly advantageous (Plate 14, colour section).

Although a few specially equipped aircraft are maintained worldwide for certain key applications (for example, hurricane and severe weather monitoring in some tropical and subtropical regions) and for ice monitoring in high latitudes, the cost factor is a very effective limit on the use of aircraft in hazard monitoring as a whole.

There are two different types of approach to the use of satellites for natural hazard monitoring:

1. the *regional* approach, exemplified by Table 18.3.
2. the *thematic* approach, as illustrated by Table 18.4.

Amongst the more common natural hazards, *earthquakes* and *volcanic eruptions* are, at least potentially, as dramatic as any in their surface impacts. Of these, the former do not readily lend themselves to monitoring by remote sensing, but the latter do. Ash clouds, e.g. from Mount Pinatubo in the Philippines in 1991, have been monitored on NOAA satellite imagery, and lava flows and environmental damage have been mapped and assessed after volcanic episodes have ended in many cases, e.g. in the case of Mount St Helens in the north-west USA after its spectacular eruptions in 1980. However, the majority of natural hazards are not of geological, but of *atmospheric*,

origin. Fortunately these lend themselves rather readily to satellite remote sensing. Consequently, we will focus much of our attention in the remainder of this chapter on the use of satellites in monitoring hazards directly or indirectly related to the state and behaviour of the atmosphere.

Authorities in the hazard monitoring field have established that the most frequent natural disasters are associated with *flooding*, whether slow or sudden in developing. Some 60% of all natural disasters are said to be of this type. On the other hand, the most frequent stimuli of disastrous loss of life are *severe tropical storms and hurricanes* (alternatively known as tropical cyclones or typhoons). In view of the significance of such events, and the detailed attention which they have been accorded in remote sensing circles, we may profitably consider severe convective storms (the instigators of most flash floods) and hurricanes in greatest detail, to illustrate the value of satellites in hazard monitoring. Our third case study will involve *insect infestations*, for interesting and varied research has been undertaken and is in progress in respect to several insect pests. Globally these are reckoned to destroy or damage one-third of all food crop production, besides inflicting many diseases on human beings and their livestock. Fortunately our ability to monitor and control some large insect populations has already been improved significantly using satellite data.

18.4 SEVERE CONVECTIVE STORMS

Certain parts of the world are susceptible to severe convective storms, which can cause considerable loss of life and damage to property through associated high wind speeds, short duration high-intensity rains and even tornadoes – locally the most destructive atmospheric phenomena on the Earth, judged by the fury of their impact in relation to their size. The USA, open in the summer half of the year to very warm, moist, convectively unstable air

Table 18.3 Hazards in north-eastern USA, which are routinely assessed by US Weather Bureau offices from the evidence of geostationary satellite images (After Wasserman, 1977; from Barrett and Hamilton, 1980)

Type of hazard	Applicability of satellite data analysis	Key satellite evidence or indications
1. Heavy rain	Direct	Significant, slow-moving convective activity within 100 km of area of interest, or upslope flow of moisture of marine origin
2. Flash flooding	Indirect	Heavy rain, as in (1)
3. River flooding	Indirect	Heavy rain as in (1)
4. Heavy snow	Direct	Significant precipitation within 100 km of area of interest, and/or upslope flow of moisture of marine origin Areas of heaviest snow indicated by imagery
5. Freezing precipitation	Direct	Liquid precipitation in or approaching the area of interest if the surface temperature is near or below freezing, or expected to fall below freezing
6. Poor visibility (highway, aviation, marine	Direct	Fog and stratus over the area of responsibility, and possible advection of thick fog into a critical area
7. High winds (other than those associated with severe convection)	Direct	Movement and change in intensity of a storm that may produce high winds
8. Severe thunderstorms and tornadoes (high winds, hail, lightning, heavy showers)	Direct	Significant convective activity within 100 km of area of interest or a triggering mechanism approaching an unstable area
9. Low-cloud ceilings at aviation terminals	Direct	Development and change in areal coverage of low clouds
10. Strong low-level wind-shear	Direct	The presence of mountain wave clouds, or boundaries of sharp discontinuities in wind flow
11. Aircraft turbulence (clear air or mountain wave)	Indirect	Cloud features as in (10)
12. Aircraft icing	Direct	Movement of clouds associated with aircraft icing reports
13. Coastal or lake flooding	Direct	Movement and change in intensity of storms with which flooding may be associated
14. High coastal or lake waves	Direct	The presence of intense convection cells over marine areas, and movement and change in intensity of storms that may cause high waves
15. Frost, freeze and cold wave	Direct	Night-time cloud-free areas favourable for radiational cooling Clues to falling temperatures in cloud-free areas in enhanced infrared images
16. Heat wave	–	–
17. Air pollution potential	–	–
18. Fine weather potential	–	–
19. Beach erosion potential	Direct	Movement and change in intensity of a storm that could cause beach erosion

from the Gulf of Mexico, has a very significant severe-storm problem.

This is magnified by the high property values in this advanced nation. Convectional activity over the coterminous USA is of special concern to the US Weather Bureau, which makes intensive use of geostationary imagery from GOES satellites in support of conventional data for monitoring severe convective situations. Of special interest is the work of the Synoptic Analysis Branch of NOAA/NESDIS, which began supporting the Quantitative Precipitation Branch (QPB) of the National Weather Service as early as 1977 with estimates of severe storm precipitation derived from satellite data by the so-called Scofield–Oliver technique (Chapter 11). There are three particular reasons why such an approach is deemed essential in a country well equipped with surface weather-observing stations, including a good weather-radar network:

1. During storms normal communications may be interrupted, and/or conventional

Table 18.4 Examples of agricultural hazards which are amenable to monitoring from satellites, classified in terms of the time periods required for them to become effective (After Howard *et al.*, 1978)

Class of hazard	Types of hazard
Short-term impact (hours to weeks)	Severe storms and hurricanes Flash-floods Forest fires Wind Frost Heatwaves Hailstorms
Short-term cumulative impact (weeks to years)	Flooding from prolonged rainfall Drought Ice and snow cover Adverse synoptic weather Persistent cloud cover Insect infestations Disease
Long-term impact (years to centuries)	Climatic change Soil degradation Desertification Pollution

weather stations damaged, so that the normal *in situ* data flow is impaired.

2. Satellites provide a more complete areal view than the conventional observing network can give. This is often of special significance in short-term forecasting of severe weather activity.

3. Much of the USA is susceptible to flash-flooding if high-intensity rains occur. Since the exact location and path of movement of a severe convective storm may be very significant in terms of the river basins it may affect, the conventional station network is often not close enough to give the spatial detail desired in such cases.

An appreciation of the scale of the flash-flood threat can be gained from the following examples, all of which occurred during one period of a little over 12 months, in 1976–1977:

1. The Big Thompson Canyon flood in Colorado, 31 July 1976: 305 mm (12 in) of rainfall, in parts in only 4 h, caused 139 deaths and an estimated US$35 million damage.

2. The Johnstown (Pennsylvania) flood, 19–20 July 1977: 305 mm (12 in) of rainfall in 10 h led to 77 deaths and some US$200 million damage. The same community had previously suffered from one of the worst flash-floods in US history: 220 known deaths in 1889.

3. The Kansas City flood, 12 September 1977: 203 mm (8 in) of rain in 6 h resulted in 25 deaths, and an estimated damage to property of US$90 million.

Although originally designed as a manual method, nowadays the Scofield–Oliver technique involves a person–machine mix on the purpose-built IFFA (Interactive Flash-flood Analysis) system, and a fully automated version is under development. Through this, visible, infrared, water vapour channel and enhanced infrared images are analysed via a selected decision-tree interpretation scheme, to establish suitable rainfall estimates for

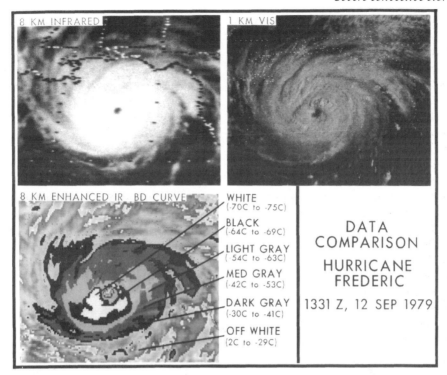

Fig. 18.3 Comparison of raw infrared and visible images of Hurricane Frederic from an American geostationary satellite at 1331 GMT, 12 September 1979, with an enhanced infrared product in which radiation brightnesses have been translated into temperatures of radiating surfaces. These can be interpreted in terms of heights above sea-level (colder cloud tops have higher radiating tops). Enhanced products are often physically more informative than unenhanced images. (Courtesy, NOAA.)

sensitive areas, assessed on a small grid-square basis (down to a few square kilometres). Of special significance is the enhancement process (Fig. 18.3), which contours cold convective cloud tops in terms of temperature (or height). Warmer portions of the imaged areas are displayed by a nearly linear relationship between radiation temperature and picture brightness (dark for warm, white for cold), up to the edges of cumulonimbus anvils (at about −30°C). Within the anvils lower temperatures are represented by a further set of grey shades from black to white. Reduced to its essentials, the Scofield–Oliver method is comprised of three stages. In these the analyst seeks to:

1. identify the *active portions* of each powerful convective system;

2. generate an *initial rainfall estimate* from the indications of the enhanced infrared imagery, plus knowledge of the typical performance of such convective systems locally;

3. make allowance for *particularly heavy rainfall* through the identification of synoptic or subsynoptic 'amplifiers'. This involves examining the life history of the event in question, and searching for evidence of:
   (a) quasi-stationary cumulonimbus systems, and/or such systems regenerating *in situ*;
   (b) cold, rapidly expanding anvils;
   (c) cumulonimbus mergers;
   (d) merging lines of cumulonimbus;
   (e) overshooting tops (rapid upward growths of cumulonimbus towers through cumulonimbus anvils).

Since the original Scofield–Oliver technique was formulated, much effort and experience also has been put into developing and improving it, and most particularly into broadening the general approach to types of storms other than summertime convective outbreaks. Today distinctive variants of the technique are available for high-intensity rains associated with winter convective storms, winter (snow producing) mid-latitude depressions, tropical cyclones (hurricanes), and 'warm top' convective storms from which heavy rains may fall despite an absence of the very low cloud-top temperatures characteristically associated with the heavier rainfalls.

Although this type of approach is complex, and even when implemented on an interactive computer system depends heavily on the skill of the analyst, its operational use in the USA has yielded many valuable forecasts in potential severe weather flash-flood situations. Simpler versions have been developed for use by the hydrological offices of other countries, e.g. in Central and South America (Fig. 18.4), whilst under the auspices of the WMO widespread transfer of the full NOAA/NESDIS technique and technology is now being actively encouraged.

18.5 HURRICANES

In the previous section we considered the use of weather satellite data in the monitoring of severe weather, which is organized characteristically at the meso-scale (10–100 km in diameter). Here we consider much larger, more strongly organized and more severe macro-scale weather systems capable of inflicting great damage to human life and property, not only *directly* through high intensities of wind

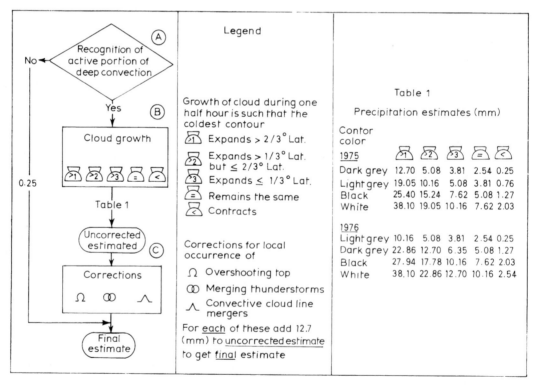

Fig. 18.4 Flow chart and table for the estimation of point rainfall from geosynchronous satellite imagery for South America. (Source: Ingraham and Amorocho, 1977.)

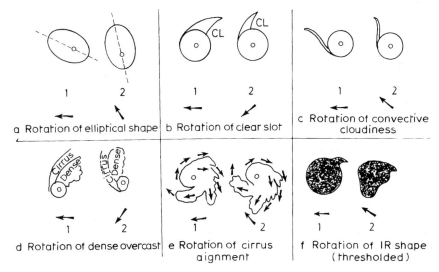

Fig. 18.5 Characteristic types of cloud patterns associated with directional changes of motion of tropical cyclones. (Source: WMO, 1979.)

and rain, but also *indirectly* through associated surface effects, especially in low-lying coastal zones. Synoptic-scale revolving or rotating tropical vortices (Fig. 18.2) – variously known as tropical cyclones, hurricanes, or typhoons in different regions of the world – are legendary for their destructive potential. Adjudged by the scale and breadth of the impact they often inflict on some tropical and subtropical lands, this is the most destructive type of weather phenomenon on Earth. Because the hurricane is so significant a threat to human life and property, but also so clearly defined in terms of its vortex of dense, organized convective cloud, the first weather satellites were conceived largely as much-needed monitors of these storms: most of their lives are spent over remote oceans and seas, whence supplies of conventional weather data are very sparse. The monitoring of hurricanes from satellites has been so overwhelmingly successful that no significant tropical storms or hurricanes have gone unseen since the first operational polar-orbiting weather satellite system was inaugurated by two ESSA satellites in February 1966. Concerted efforts have been made, especially in the USA, to interpret weather satellite imagery as fully as possible to improve both the monitoring of existing

hurricanes and the forecasting of their likely movements and changes in the future. Currently, satellite imagery is used to:

1. monitor hurricane *tracks*;
2. assess hurricane *intensities*, and estimate their *maximum wind speeds*;
3. identify *areas of influence* of each hurricane;
4. estimate associated *volumes of rain*, although schemes for this are the most tentative of the four.

It is instructive to consider each of these in more detail.

18.5.1 HURRICANE TRACKS

In some cases it is easy to identify the centre of a hurricane because a well-defined, relatively cloud-free 'eye' is plainly evident in or near the middle of the dense clouds of the vortex (Fig. 18.2). However, on many occasions, especially in sub- or post-mature situations, centre location is difficult because no eye is visible and/or the vortical cloud is fragmentary in appearance. Here it is necessary to distinguish between the *central features* (CF) of the hurricane cloud system and peripheral *banding features* (BF), which are arcs of cloud tending to point or spiral towards the centre of the storm circulation in the lower troposphere.

Through careful tracing of cloud arcs, a convergence centre can be established with reasonable confidence in many cases. Results compiled from many hurricane seasons show that tracking is usually correct to within $\frac{1}{4} - \frac{1}{2}°$ (28–55 km) for the location of the centre of each storm. Only occasionally – in respect of particularly ill-organized storms – are centre locations 1–2° or more astray. For forecasting future movements of the storms, recognition of significant changes of vortical cloudiness is often profitable (Fig. 18.5).

18.5.2 HURRICANE INTENSITIES

Early classifications of tropical storms and hurricanes revealed that, through considerations of their sizes and stages of development, good estimates could be made of sustained maximum wind speeds (MWS). This work led to the development of highly detailed classification schemes, which can be used both diagnostically and predictively. This is illustrated by Fig. 18.6, which categorizes common tropical cyclone patterns in terms of 'T-numbers' that are largely (but not entirely) determined by the sum of values allotted to CF and BF characteristics.

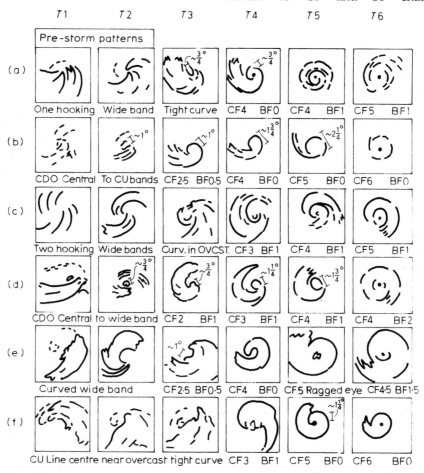

Fig. 18.6 Common tropical patterns and their corresponding 'T-numbers'. The T-number shown must be adjusted for cloud systems displaying unusual size. The patterns may be rotated to fit particular cyclone pictures. (Source: WMO, 1977.)

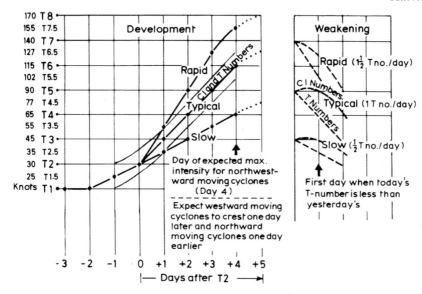

Fig. 18.7 Current intensity change curves of the tropical cyclone. The shaded area surrounding the typical curve represents 'intensity' as a zone, one T-number in width. (Source: WMO, 1977.)

Comparisons with surface and aircraft observations have shown that current intensity (CI) numbers (T-numbers modified to take account of influential meteorological factors not directly evidenced by the cloud features alone) can be interpreted diagnostically in terms of MWS, as exemplified by Fig. 18.7. The T-numbers alone can then be used predictively to give forecasts of hurricane wind speeds for up to a few days ahead, once it has become clear whether the storm in question is following a rapid, normal or slow development curve. Verification tests have indicated that assessments of hurricane intensities expressed in terms of MWS are 90% correct within two T-number categories (i.e. ±1.0T) which, translated into wind speeds, is ±15 knots. This is reasonable, remembering that hurricane winds often substantially exceed 100 knots.

18.5.3 AREAS OF HURRICANE INFLUENCE

Here the aim is to establish the corridor likely to be affected by strong winds and heavy rain, especially as a hurricane makes a landfall and tracks inland. For this purpose attention is

paid to the central dense overcast (CDO) in conjunction with the predicted path of the storm. The CDO is of greatest significance for likely rainfall, but banding features also may have to be taken into account in order to delimit the total regions of possible wind damage.

18.5.4 HURRICANE RAINFALL

This has been particularly difficult to estimate accurately, owing both to the extreme rainfall intensities that hurricanes may bring and the propensity for hurricane rainfall to vary spatially to a high degree. Experience with infrared images of hurricanes in the American sectors of the North Atlantic and North Pacific Oceans has indicated that the following rainfall rates may be assumed as representative of a transect marked by the storm centre path:

1. *Wall cloud area* (around the eye, within 20 nautical miles of the storm centre): 51 mm h^{-1} (±25 mm).
2. *Inner central dense overcast* (CDO) area (within 50 nautical miles radius of the storm

centre, adjustable on the evidence of enhanced infrared imagery): 25 mm h^{-1} (±12 mm).

3. *Outer CDO* area (radius defined by the outer edge of the CDO): 1–2 mm h^{-1}. If embedded convective bands are present these rates are expected to be higher, between those of 2 and 3.

At the time of writing efforts are being made to estimate instantaneous rainfall rates for hurricanes more confidently by using passive microwave image data from the US military satellite family, DMSP (Chapters 5 and 10), and from the TRMM satellite. The early results are promising, not least because these data reveal much more directly the areas of rainfall embedded in the CDO and BF clouds, as shown by Fig. 18.8 and Plate 7 (colour

section). They are also thought provoking, because most of the storms so far analysed in this way seem to have had rainfall patterns different from those that existing hurricane models (largely based on visible and infrared imagery) suggested, emphasizing the *need to treat each such storm, and each storm stage, individually*.

Calibration of the passive microwave algorithm outputs of rainfall rates is proceeding, but slowly, because radar data are required for this purpose, and severe hurricane occurrences in areas served by suitable (digital) radars are rather infrequent. A further challenge stems from the fact that passive microwave data are only available at present from low-orbiting satellites, which cannot provide the temporal detail that fast-changing hurricanes sometimes demand; thus one of today's

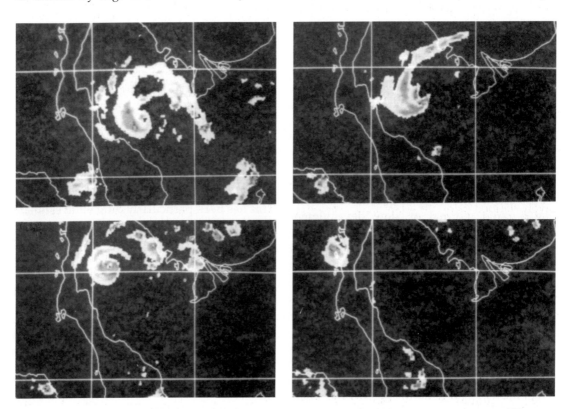

Fig. 18.8 DMSP – SSM/1 passive microwave rainfall intensity images of typhoon Gay over the Gulf of Thailand each day from 1–4 November 1989. (Courtesy, Dr. S. Huntrakul.)

key research problems is how best to combine the physically more direct (but spatially less detailed and temporally less frequent) evidence of rainfall afforded by the low-orbit passive microwave data with the physically less direct (but having higher spatial and temporal resolution) infrared data from the high-orbit geostationary satellites.

However, we must conclude that satellites are vital tools for use in hurricane monitoring and forecasting generally, remembering that this use of satellite data alone has justified, and will continue to justify, the operation of environmental satellite systems for many countries in the tropical and subtropical world.

18.6 INSECT INFESTATIONS

Insects are highly significant to the human race, for, as we have seen, it has been estimated that much of the annual food production of the world is rendered inedible because of damage by insects. Insects are significant to us also because of the many diseases which they transmit, or more directly cause.

Researchers in remote sensing have long pursued the possibility that conditions conducive to upsurges of several major insect pests might be monitored from aircraft. Perhaps the best examples are from the mosquito family, whose breeding habitats have been extensively defined using multiband aerial photography. More recently, attention to the potential of satellite remote sensing for improved pest monitoring and control has broadened greatly, although the apparent promise of satellites for such purposes has, perhaps, still to be generally fulfilled. In principle at least this work has involved a variety of insect pests, influenced by various environmental parameters and characteristics. Here we will consider in more detail projects relating to three pests which seem especially amenable to monitoring from satellite altitudes, and discuss ways in which operational success in respect of each might be achieved.

18.6.1 THE SCREWWORM (*COCHLIOMYIA HOMINOVORAX*)

This is a devastating cattle pest of Central America and southern North America, being endemic in Mexico, and expanding into the USA in summer. Its adult stage is temperature controlled, the fly being especially sensitive to both very hot and cold weather (for this reason there are no over wintering populations in the USA). The distinctive features of the species are that:

1. the larvae feed only on living tissue;
2. the population density is relatively small;
3. the female flies mate only once in their lifetime.

Point 3 renders the fly vulnerable to the introduction of sterile males, which therefore can be used as an effective control measure. Without this, annual losses in the USA could exceed some US$600 million.

In an experimental Screwworm Eradication Program of NASA, high-resolution infrared data from NOAA satellites were processed in conjunction with correction factors for atmospheric attenuation to give daily maps of air temperatures near the ground in northern Mexico and the southern USA, registered to a common geographic grid. Longer-term temperature maps and maps of degree days were also produced. These were judged to have been successful, sterile flies having been released by aircraft over significant areas of screwworm activity. Unfortunately, as project staff reported, 'The logistics of handling images on computer tapes turned out to be a major problem, and the extensive processing required to digitize, register, calibrate, and produce the various products was expensive.' For such a scheme to be implemented, further model and programme development seems necessary, especially in developing countries where the need is greatest – computer costs, however, have fallen dramatically. With the recent spread of the screwworm to Libya in North Africa (in which continent it was previously unknown) the need for successful control measures is even greater than before.

18.6.2 THE DESERT LOCUST (*SCHISTOCERCA GREGARIA*)

The desert locust is both a traditional scourge of, and a continuing threat to, the agricultural economies of many countries stretching across the desert and semi-desert belt of the Old World, from western Africa through the Middle East to south central Asia. To counter and contain this threat a multinational organization, the Desert Locust Commission, has been set up. This meets annually in the headquarters of FAO in Rome, which helps to coordinate the work of the area Desert Locust Control Commissions, and organizes research into new methods of survey and control. In particular, detailed satellite research began in 1975, focused initially on the area of the North-West African DLCC (covering Algeria, Libya, Morocco and Tunisia), with the aim of developing an operational method for satellite-improved desert locust monitoring and control (Fig. 18.9).

The desert locust is ideally suited to the exploitation of favourable conditions for

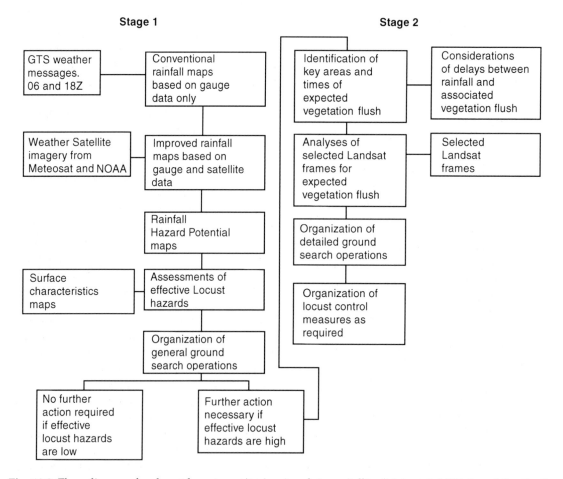

Fig. 18.9 Flow diagram for desert locust monitoring involving satellite (Meteosat, NOAA and Landsat) imagery. This scheme was designed for the Desert Locust Commission and FAO, beginning with North-West Africa.

breeding as and when they arise. In periods of adversity there is a basal population of locusts spread thinly throughout the affected zone. Some of these 'solitary locusts' become concentrated in rain-fed areas, due largely to the vacuuming effect of converging winds. Here, the solitary locusts change their shape, colour and pattern of behaviour as their feared 'gregarious phase' begins.

The desert locust is capable of considerable migratory movement in search of food and breeding grounds. In suitable areas its reproductive rate achieves geometrical proportions. Suitable conditions for breeding include:

1. a sandy-silty soil for easy egg-laying;
2. sufficient soil moisture in the top layer of the soil to encourage eggs to hatch;
3. green vegetation for the immature locusts to feed on after hatching.

Since 2 and 3 are determined principally by the distribution of rainfall events, *improved monitoring of rainfall* is a primary objective of a satellite-assisted desert locust control operation. However, 1 is determined by desert morphology and pedology and both 2 and 3 are also influenced by surface characteristics. Consequently Landsat and SPOT imagery have been used for the compilation of *surface characteristic maps* of locust-affected areas, which are much more detailed than existing topographic maps.

In the FAO/DLC research scheme of the late 1970s and early 1980s (Fig. 18.9) weather satellite imagery was used along with surface weather station data to provide the best possible maps of rainfall for periods of 12 h and upwards. Such maps are interpretable in terms of related 'locust hazard levels', for two important practical applications:

1. The identification of areas where significant rain is thought to have fallen in areas whose surface characteristics might be conducive to vegetation flush and locust breeding; post-rainfall Earth resources satellite imagery from such areas can be analysed to reveal the extent and density of any significant growth of vegetation. This should then be inspected in more detail from the air or on the ground.
2. The planning of routes to be followed by ground inspection teams, so that these can be directed more efficiently and economically than before to areas of greatest locust breeding hazard.

Results obtained from the operational research programme in its North-West African test area were very encouraging. Since vegetation flush post-dates rainfall by periods from a few hours in spring to weeks or months in autumn and winter, and the locust life cycle from egg-laying to the first flight of a new generation spreads over about 21 days, there is ample time for the collection, analysis and interpretation of the satellite data and the planning of detailed 'search and destroy' operations before local population dynamics could become critical for the development of plagues of the desert locust.

The project summarized above is a good model of ways in which remote sensing and *in situ* observations can be combined for greatest possible effectiveness, efficiency and economy of pest monitoring and control programmes, and has since been built upon for present operational strategies. However, modifications to the original approach were necessary before an operational version could be implemented, primarily because of the high costs of obtaining and analysing data from the Earth resources satellites, Landsat or SPOT. Fortunately, significant cost savings in implementing such an approach now can be made because of the serendipitous discovery in the early 1980s that data from bands 1 and 2 on the NOAA–AVHRR sensor could be ratioed to provide useful 'vegetation indices', as explained in Chapters 13, 15, and 16. These indices, which provide evidence of *photosynthetic activity* – and even, in areas of grasses and other perennial plants, of *biomass*, and its seasonal changes – can be generated weekly at spatial resolutions down to nominally 1 km.

aids we have in detecting hazardous weather in its earliest stages of development'. Even by that early date, set procedures existed for the duty forecaster to consider the value of the available satellite data in the preparation of a wide range of special advisories or 'nowcasts' to warn of current or expected weather and weather-related hazards (Table 18.2).

More specialized, but perhaps even more important, nowcasts and warnings based largely on satellite data analyses are issued by offices such as the Hurricane Forecasting Center in Miami, Florida, and an increasing number of other national weather offices in areas susceptible to severe weather.

Increasingly, both environmental and Earth resources satellite data are being used in a widening range of other environmental contexts too, especially related to agricultural activity. *Crop hazards* can all too easily lead to shortages of food, increased food prices, and even famine. Longer-term results of satellite monitoring of crop areas and yields (discussed in Chapter 16) may even involve changes in the agricultural policies or internal and external politics of nations or groups of nations.

The Food and Agriculture Organization of the United Nations, with its headquarters in Rome, is vitally concerned with the global availability of food crops, and crop shortages: one of its briefs is to improve 'Food Crop Security', i.e. to help ensure 'stability of an adequate food supply for all.' Unfortunately the intelligence required for international technical assistance and relief activities is often not available from conventional (ground) sources as quickly as it needs to be if major human disasters are to be averted. However, the satellite, equipped to observe changing conditions and events frequently and very widely over the surface of the Earth, is helping greatly, both in the forecasting of crop harvest volumes and the detection and monitoring of acute hazards that are threatening to agricultural practices and production. The FAO has sought to respond to this challenge, especially through its ambitious ARTEMIS system. The value of

such programmes is recognized also by the commercial world, for which a growing number of meteorological bureaux and consultancy firms generate – and frequently update – information on the areas and conditions of major crops.

Even at smaller scales, including those of local regions, cities, towns, rural districts, rivers and coastal waters, increasing uses are being made of satellite data to improve hazard monitoring as 'distributed systems' of satellite data collection and analysis become possible. Below these scales, and for key instances in time, airborne and even seaborne systems come into play more and more to evaluate disasters and assist with relief operations.

Certainly there is still much unexploited potential in remote sensing, not only for the more efficient and economic monitoring of environmental hazards but also for the increased avoidance or mitigation of natural disasters. What is needed now is a concerted research effort to establish the optimum techniques of data analysis with such ends in mind, and a strengthening of those organizational structures that would foster such approaches, in addition to, or in place of, the existing procedures where these are inadequate to meet the present needs.

Thus, broad-scale improvements in flood forecasting and mitigation schemes, such as those envisaged in Fig. 18.11, require advances not only in scientific knowledge and understanding but also in systems engineering, and above all, in international relations even in spheres not previously noted for cross-border cooperation.

Certainly, hazard monitoring and disaster prevention and mitigation increasingly demand action on continental or even global scales. Perhaps the two events in recent years which, more than any others, have focused the attention of the world community on the needs for satellite monitoring of hazards and evaluations of associated disasters, and the present opportunities for an associated greatly improved watch on the world at all scales from

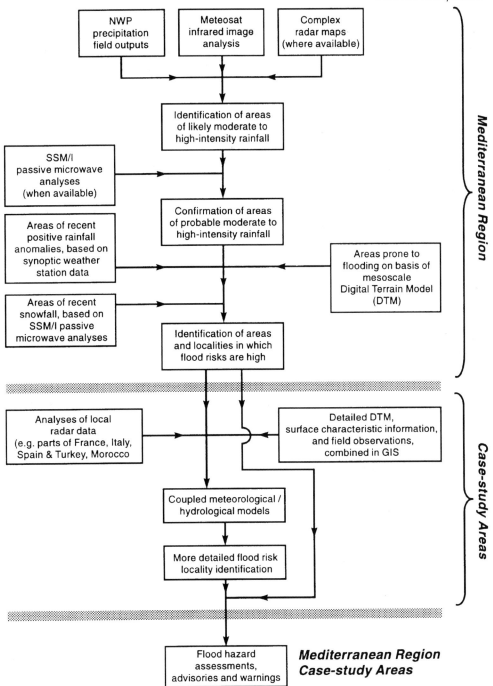

Fig. 18.11 A proposed approach to flood monitoring and forecasting in the Mediterranean region, using inputs from several and varied sources. (Figure: E. C. Barrett, CRS, University of Bristol.)

encouraged by the major satellite operators. Ground stations for the capture of data from geostationary weather satellites (e.g. Meteosat) are particularly cheap. Reception facilities for some near polar-orbiting satellites (e.g. those of the NOAA family) are also widespread. In this respect, advances in ground receiving station design are now proceeding rapidly, and polar-orbiting satellite receivers are becoming substantially cheaper than before, especially those using 'fixed horn' or manually operated antennae instead of mobile automatic 'tracking' antennae.

As ground station designs improve further, and associated facilities become more easily affordable than before, other constraints on the widespread use of satellite data need to be relaxed – for example, the present encryption of signals from some (civilian as well as military) environmental satellites (e.g. Meteosat), and the use of narrow bandwidth transmitters by which the data are transferred from some other satellites to the ground. Meanwhile, as suggested above, access to Earth resources satellite data has been almost uniformly restricted by the insistence of the satellite operators to issue few ground station licences, partly because of the security implications of their much higher resolution data, but mostly because of the *higher market values* of their images and the greater (though still elusive) hopes held out by the operators for commercial profits from their space business.

So far as security issues are concerned, there are several apparent sources of difficulty. For example, many sensors and sensing analysis and interpretation systems are classified as 'secret', and are therefore available to military

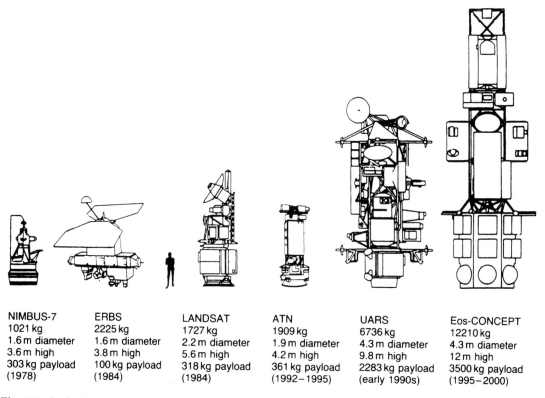

NIMBUS-7	ERBS	LANDSAT	ATN	UARS	Eos-CONCEPT
1021 kg	2225 kg	1727 kg	1909 kg	6736 kg	12210 kg
1.6 m diameter	1.6 m diameter	2.2 m diameter	1.9 m diameter	4.3 m diameter	4.3 m diameter
3.6 m high	3.8 m high	5.6 m high	4.2 m high	9.8 m high	12 m high
303 kg payload	100 kg payload	318 kg payload	361 kg payload	2283 kg payload	3500 kg payload
(1978)	(1984)	(1984)	(1992–1995)	(early 1990s)	(1995–2000)

Fig. 19.1 Scale drawing comparing one concept for an EOS payload on a platform with current-generation spacecraft, illustrating stages in the evolution of environmental satellite sizes, complexities and payloads. (Courtesy, NASA.)

users only. In some cases security classification rulings have seriously inhibited the exploitation of new satellites or sensors for peaceful purposes. The following are some of the ways in which such restrictions have become evident:

1. Through restrictions on the availability of data and equipment for present programmes.
2. Through restrictions on the education of scientific and engineering personnel who could use new tools and new data with considerable benefits in civilian application areas.
3. Through obstructions to the formulation and development of similar research in the civilian domain.
4. Through the erection of barriers between scientists and engineers concerned with application areas that demand higher resolution data (spatial, spectral or temporal) than those presently available to the civilian user community.
5. Through limitations on the free flow of technical information within the world community, often resulting in undesirable duplication of effort and incomplete reporting of important research experience.
6. Through artificial restrictions on the development of civilian markets so that system and component costs remain artificially high.

In these areas of difficulty there is clearly a problem of balancing potential gains to the civilian community against potential perceived losses in respect of national and international security. The related judgements are never easy to make, but it is clear that the classification of any remote sensing devices and data as secret is now rapidly becoming more and more difficult both to defend and enforce.

Implicit in these questions concerning the accessibility – or otherwise – of data from satellite remote sensing platforms is the administrative and legal problem of whether 'open sky' observations are permissible or not. For many years there have been differences expressed in the Outer Space Committee (OSC) of the United Nations as to a variety of difficult issues of international law, including:

1. whether countries have the right to obtain remote sensing imagery of or from any other country;
2. with whom ownership of the satellite imagery obtained over a particular country resides;
3. whether imagery over a particular country may be made available to neighbouring – perhaps unfriendly – countries against the wishes of the first.

Fortunately for those who believe that open access to all satellite data is desirable, not only are data from Western satellites available with a very high degree of freedom, but increasingly access to imagery obtained by the former Soviet Union has recently become much more open than before. Perhaps understandably though, most countries restrict access to high-resolution imagery over specially sensitive areas. Whether this still makes sense in an era in which intelligence services have very sophisticated and varied means of obtaining 'target' information from satellites is increasingly open to question. In any case, the increasing availability of sub-10 m resolution satellite data from civilian sources is forcing the issue.

19.2 REMOTE SENSING AS A PUBLIC SERVICE OR A COMMERCIAL ACTIVITY

Relating to the discussion in the preceding section is another increasingly important question, as to whether satellite remote sensing should be primarily a *public service*, or a *commercial activity*. During the last decade there have been concerted attempts by some governments, notably of the USA and France, to ensure that the costs of national satellite remote sensing programmes should involve the private as well as the public sector. Fortunately or unfortunately, such efforts have only met with limited success. In the USA it is now recognized that considerable public subsidies must be made available to ensure that pivotal programmes such as Landsat will continue, whilst recognizing that there is some role for commercial companies in respect of the marketing of data and 'adding value' to it via customer-oriented products. In France the experience has been broadly similar, although a stricter policy has been applied to the marketing of SPOT data than has been the case with Landsat.

A further issue relating to all of the above is that transfer of remote sensing to the private sector may negatively affect the benefits of satellite remote sensing to the world community as a whole, not least because developing countries stand to benefit most from a situation within which relatively low-cost access is possible to data from space, since in this case the primary costs of establishing the remote sensing systems have been borne by others, without being passed on to the end-user. More broadly it is clear that relatively open access to data from space is much more certain to encourage a continuing increase in the rate of penetration of applied remote sensing into a wide range of areas and monitoring programmes, whereas increased commercialization of space may at least temporarily slow down the rate of such growth. Section 19.6 is important to this debate.

19.3 HANDLING LARGE DATA SETS

Great problems are encountered today in the field of data processing as a result of the truly vast and increasing quantities of data that can be gathered by modern remote sensing systems. For example, the Landsat MSS produced 15 megabits per second of data, whilst the Thematic Mapper generates over 10^4 megabits per second, and much higher data rates may be expected soon from the larger space platforms of the Earth Observation System. The acquisition and processing of digital image data therefore demands very significant, and increasingly large, computer resources. For example, a user of Earth observation satellite data who wishes to undertake pans, zooms, three-dimensional displays and movie-loops of satellite data requires a workstation of at least 20 MIPS performance, with some other applications requiring much higher figures even than these. In such cases memory management problems are also increasing, especially because data compression algorithms to reduce memory requirements are not very readily available.

Fortunately there have been rapid advances in electronic computing since the first Earth observation satellite was launched in 1960, and this is an accelerating trend. Significant increases have been achieved recently in both the capacity and operating speeds of computers suitable for handling remote sensing data sets (see Chapter 1), yet at the same time unit computing costs have come down quite rapidly. A very recent trend of particular significance in these respects is the development of efficient software packages that free the analyst from the earlier requirements for specialized hardware-dependent systems. At the same time, there has been a trend towards the development of *distributed computer networks* through which operators at a number of workstations can function simultaneously on a particular data set, whereas previously only one might have been able to function at a time. However, there are still difficulties in sending large images and attendant voluminous data sets around such networks, owing to limited bandwidths. Developments in communications technology will be required to alleviate this kind of restriction. An associated problem is the increasing trend of problem 'multi-sourcing', i.e. involving data from many different sources of data (both *in situ* and remote sensing) for best overall effects. Whilst eminently sensible and praiseworthy, this further exacerbates problems of data volumes and data handling.

In Chapters 1, 7 and elsewhere in this book it has been noted that an important development in the last decade has been in the growth of Geographic Information Systems, through which many different types of spatially distributed sets of data may be organized and intercompared. Today, remote sensing data inputs are seen as increasingly important for modern GIS, and it is certain that the GIS of the future will increasingly capitalize upon the more dynamic types of data which satellite remote sensing systems – both Earth resources and environmental – are able to provide.

In all these areas, however, there have been growing complaints from the remote sensing user community concerning the *plethora of available media* for data analysis and display, the data formats, and the software packages available for data analysis and interpretation: incompatibility between systems in different laboratories is much more common than the compatibility that we need for increased cooperation within the remote sensing community. Certainly it is increasingly necessary for standards to be agreed so that systems can become more compatible with each cther, and so that the results and processing packages from different sources can be much more readily interchanged.

19.4 ARCHIVING AND DISTRIBUTION PROBLEMS

Of course, the wish to handle large data sets does not always or only relate to data that have been obtained directly from the satellites themselves, but often in relation to such data that have been obtained by some central facility. Issues related to this 'second-hand' mode of operation include *archiving* and *distribution*. The long-term storage of data and their retrieval is a considerable problem, for in many areas inadequate funds have been made available to ensure that all, or even a significant proportion of, the original data are kept for posterity. Unless this problem is addressed head on, it is one that will get much worse with the advent of the new higher data rate sensors expected in the mid-term future. *Data reduction* has been considered a high priority in some areas as an alternative to archiving an entire satellite data set. However, if data reduction is necessary it is important that any transformations leading to derived parameters prepared for archiving are reversible; otherwise, no subsequent analysis is possible of data at the earlier, more complete levels of detail.

In the related area of data distribution, recent developments have afforded increasing hope for the future. In the past most satellite data have been stored on computer-compatible tapes. These are known to suffer from a number of limitations, including:

1. limited lifetimes (theoretically as short as about 1 year);
2. weight and bulkiness;
3. unsuitability for on-line storage;
4. lack of speed in accessing;
5. differences and inconsistencies in the formats of the included data.

In several of these respects it is fortunate that the storage and use of data on exabyte and DAT tapes and on optical discs now proffers significant improvements on the above, although at present there are greater risks associated with them, for no standard sizes or formats have yet been agreed.

Perhaps the most exciting new development that is taking place in this field concerns the development of 'intelligent data management systems' (Fig. 19.2), through which users will be able to interact with even the most complex database systems, with minimal understanding of the architecture of the system, the stored data, or the query language. Thus, even as data sets become larger, more numerous and more complicated, data selection and manipulation by the research worker or commercial user should become less difficult than it has been hitherto.

19.5 RELATIONS BETWEEN TECHNOLOGIST AND USER

Following on from section 19.4, a major problem area at the present time is that of the *interface between remote sensing technologists on the one hand and the users of remote sensing data on the other.* Defining the term 'user' is somewhat difficult in practice because there are many users with different objectives. Broadly speaking, however, we may say that there are two main groups of users, the academic and the operational. In the first category one can include Earth scientists in universities and other research institutes. The second category of users are typified by National Parks, urban planning authorities, agricultural

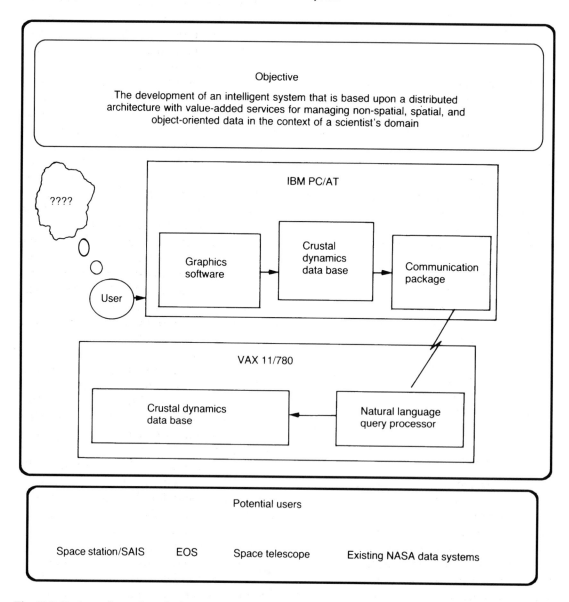

Fig. 19.2 Basic configuration of a first-generation prototype intelligent user interface. (Source: NSSDC, 1989.)

departments, meteorological services, water authorities, geological surveys and mineral companies. Each of these groups of users and others finds difficulty to a greater or lesser extent in:

1. comprehending the nature and significance of remote sensing data obtained by remote sensors;
2. defining its own requirements in terms of the engineering and physical properties of sensors used in remote sensing.

Meanwhile, it is also apparent that the space technologist does not always fully comprehend the dynamics or properties of the environmental surface being sensed. Furthermore, as increasing use is made of non-visible waveband imagery rather than conventional photography, the problems of communication between the remote sensing technologist and the general user become more and more severe. To the uninstructed person an image obtained by infrared or microwave systems looks like a photograph. Its record, however (as we have seen earlier in this book), is in most cases an *entirely different information array* from that obtained by photography. For example, in microwave studies the measured response (image) obtained from a land surface is affected by complex factors, which include:

1. characteristics of the microwave sensing system (e.g. polarization direction, observation angle and frequency of bands used);
2. complex characteristics of the surface sensed (e.g. electrical and thermal properties, surface roughness size and temperature and its distribution).

In such circumstances it is clearly desirable that there should be an appropriate flow of information from the physicist/technologist to the Earth science user. However, it is equally true that many physicists/technologists have very limited knowledge of the physical mathematical models of the Earth phenomena being sensed. This has sometimes resulted in exaggerated claims being made for the usefulness

of remote sensing systems. For instance, some early writers on air photographs and soil studies led potential users to think that soil mapping could be achieved easily by air photo-interpretation. Such statements were based on a lack of knowledge of soil classification, yet were made by some whose classification techniques affected the objectives of air photo-interpretation. Lack of environmental training has further resulted in some researchers collecting second-rate data, e.g. concerning land surface conditions for comparison with highly accurate remote sensing data.

It is highly desirable that teams working in remote sensing should be of a multidisciplinary character from the outset. If the eventual users are involved at the beginning, they can help to solve many of the technical problems, such as the types of output required. Such participation also would allow the user to make an objective evaluation of how to make best use of the information (see, e.g., Plates 13 and 14, colour section). The need for such involvement by the users is underlined by this comment made in respect of crop inventory studies during an early symposium on results from Landsat: 'From the standpoint of applications almost every investigation lacked complete definitions of technique and procedures which would allow a quasi-operational project to be undertaken.'

Whilst it was true that some of the early remote sensing systems grew without due regard to the needs of the community that might be able to use them, this has become less true of more recent systems, especially those clearly related to specific end-user needs (e.g. in meteorology), and is becoming markedly less so with systems now under discussion and development for the early twenty-first century. However, this is an area which calls for continuing care and attention.

Meanwhile some potential user agencies still do not seem to recognize, understand and plan for appropriate uses of remote sensing data. It should be recognized that fruitful use of remote sensing data depends not only on

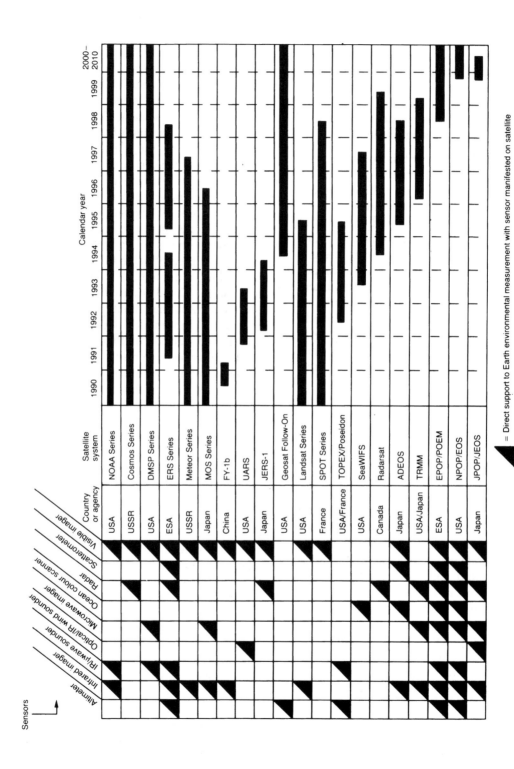

Fig. 19.3 Current and planned low-altitude satellite systems to support Earth system science, 1990 to 2110. (Courtesy, NOAA.)

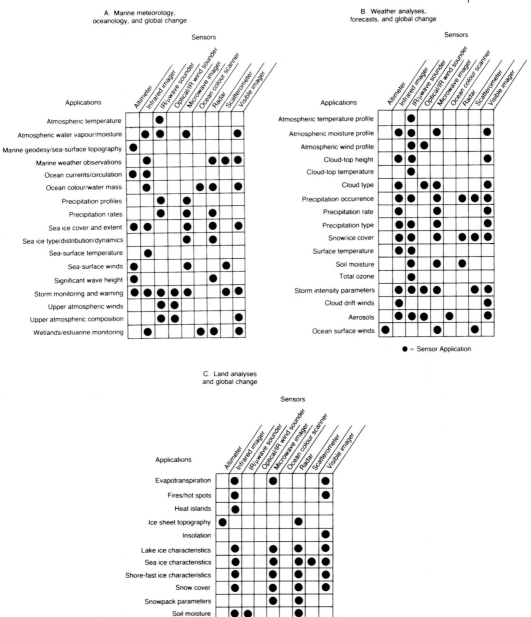

Fig. 19.4 Satellite-derived data expected for environmental monitoring applications, 1990 to 2010. (Courtesy, NOAA.)

world (Fig. 19.5). Envisaged originally as a programme to analyse data from the DSMP SSM/I sensor, with special reference to precipitation, this multidisciplinary project provided a set of laboratories with common data sets and analytical tools to encourage their scientists to work much more interactively with each other than had been the case before. Through the WetNet network scientists were encouraged to exchange their thoughts, findings, algorithms and future plans quickly via computer networks so that the WetNet science group could together make far more rapid progress in the understanding of particular types of data, and their practical use, than they could ever expect to do if working in a more traditional scientific way – largely isolated, except when meeting for conferences and workshops, and more or less competitive.

Today, the scope to work in this way is now much improved, using the Internet.

Attention must now be paid, too, to the development of systems whereby the new information resulting from the collection of remote sensing data, and its analysis and interpretation alongside other data sets, can help to fill the knowledge gap which plagues effective, efficient, and economic management of terrestrial space and resources (Figure 19.6). It will be through imaginative use of this type of system that the remote sensing community of the future will be most likely to support improved monitoring of our precious terrestrial environment, and help us to manage it in the way that it, we, and our children, all deserve.

As we draw this fourth edition to a close, it seems appropriate to quote another paragraph from the end of the second and third editions:

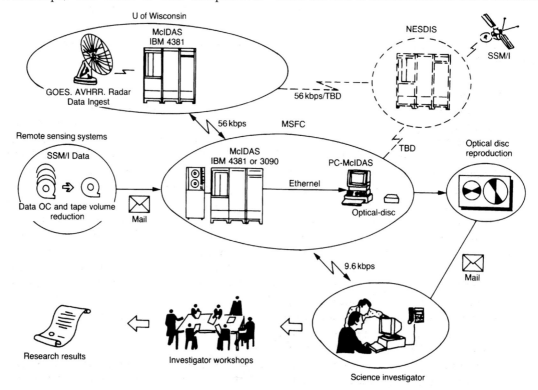

Fig. 19.5 The WetNet Project: an interactive data acquisition, dissemination, interpretation and algorithm development system. Headquartered at the Marshall Space Flight Center (MSFC), Huntsville, Alabama, this is a prototype of the EOS data systems thought likely to be necessary from early in the twenty-first century. (Courtesy, NASA.)

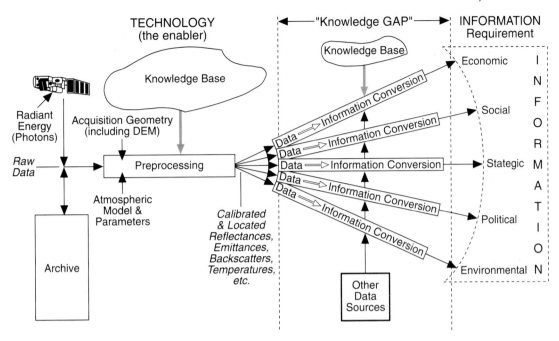

Fig. 19.6 Elements of a modern Information Delivery System to maximize use of remotely sensed data by a broad spectrum of end-users. (Source: MacDonald, 1998.)

'Although it is probably inevitable that environmental remote sensing will increasingly adopt the mantle of a 'big science' operation, it is essential that it listens to, and takes account of, the needs of the 'small' user. In recent years, the environmentalist has been involved more in satellite design than in earlier years. For example, it was claimed that Seasat was the first satellite designed primarily by the user for the user. However, it is all too easy for the leaders in the field (the 'big users') to make assumptions of the needs of others which do not match reality amongst most members of the world community of nations. The celebrated microchip should help bring the benefits of space technology within the purchasing power of many potential users less interested in multipurpose analysis systems than hard-wired 'dedicated' processors designed to give specific answers to narrowly framed questions. Beyond that, the ultimate goal for environmental remote sensing on the global scale may be a fully-automatic, general purpose, computerized satellite system. This would sense resources in several wavebands, automatically identify them, weigh them against previously programmed data on the cost effectiveness of various management possibilities, and send to the ground a decision on what should be done. It also could be used to monitor developing situations, such as a forest fire, suggesting how ground crews might fight it, and to perform automatically such tasks as turning irrigation valves on and off as required.'

Whilst such sentiments are still generally true, we must now counterbalance such thoughts with this provocative statement from Dr Lawrence W. Fritz, Secretary General of ISPRS (International Society of Photogrammetry and Remote Sensing), and

Database, and Natural Hazards Research and Applications, Information Centre):
http://www.colorado.edu/hazards/litbase/litindex.htm

IRSA (Indian National Remote Sensing Agency):
http://www.stph.net/nrsa

NASA (US National Aeronautics and Space Administration):
http://www.nasa.gov, and
http://140. 90.207.25:8080/EBB/ml/nic00.html

NOAA (US National Oceanic and Atmospheric Administration):
http://www.noaa.gov

PATHFINDER (NOAA/NASA quality-controlled research data sets):
http://www.vims.gsfc.nasa.gov/vims

RADARSAT (Part of the Space Operations Sector of the Canadian Space Agency, managing CSA's first Earth observation satellite):
http://radarsat.space.gc.ca/adrohomepage.html

RSS (Remote Sensing Society, UK: a major professional society, organizing conferences, publishing a newsletter and maintaining a useful library and website):
http://www.geog.nottingham.ac.uk/rss/index.html

SPOT Image (France):
http://www.orstom.fr/hapex/spot/spotim.html

UNEP (United Nations Environment Programme: UNEP's Global Resources Information Database (GRID) includes global and European datasets for environmental researchers, including vegetation, topography, population and climatology):
http://www.grid.unep.ch/gridhome.html

Weather Net (provides access to thousands of forecasts and images, and has nearly 400 links to North American weather sites):
http://cirrus.sprl.umich.edu/wxnet

WW2010 (a WMO World Weather Watch framework for current and archived weather data, and multimedia instructional resources using new and innovative technologies):
http://ww2010.atmos.uuvc.edu

SELECTED BIBLIOGRAPHY

Chapter 1

Abiodun, A.A. (1978), The economic implications of remote sensing from space for the developing countries, in *Earth Observation from Space and Management of Planetary Resources*, ESA SP-134, European Space Agency, Paris, 575–584.

American Society of Photogrammetry (1983), *Manual of Remote Sensing*, 2nd Edn (Vols I and II), AMS, Falls Church, VA.

Barrett, E.C. and Curtis, L.F. (eds) (1977), *Environmental Remote Sensing: Practices and Problems*, Edward Arnold, London.

Chrisman, N. (1997), *Exploring Geographic Information Systems*, John Wiley, New York.

ESA (1991), *Report of the Earth Observation User Consultation Meeting*, ESA SP-1143, European Space Agency, Paris.

European Association of Remote Sensing Laboratories (1996), *EARSeL Directory*, EARSeL, Paris, France.

Frank, A.U. (1988), Requirements for a database management system for a GIS, *Photogrammetric Engineering and Remote Sensing*, **54(11)**, 1537–1564.

Lagarde, J.B. (1980), Space and meteorology – an intricate cost/benefit ratio, in *Proceedings of the International Colloquium on Economic Effects of Space and Other Advanced Technologies*, Strasbourg, 28-30 April, ESA SP-151, European Space Agency, Paris, 175–183.

MacQuillan, A.K. and Clough, D.J. (1978), Benefits of spaceborne remote sensing for ocean surveillance, in *Earth Observation from Space and Management of Planetary Resources*, ESA SP-134, European Space Agency, Paris, 585–596.

NASA (1987), *From Pattern to Process: the Strategy of the Earth Observing System*, EOS Science Steering Committee Report Vols. I and II), National Aeronautics and Space Administration, Washington, DC.

NASA (1988), *Earth System Science: a Closer View*, Earth System Sciences Committee, NASA Advisory Council, National Aeronautics and Space Administration, Washington, DC.

National Geographic Society (1998), *National Geographic Satellite Atlas of the World* (1998), National Geographic Society, Washington, DC.

NERC (1989), Geographic Information Systems in the Environmental Sciences, *EARSeL News*, 38.

Peel, R.F., Curtis, L.F. and Barrett, E.C. (eds) (1977), *Remote Sensing of the Terrestrial Environment*, Butterworths, London.

US Congress (1985), *International Cooperation and Competition in Civilian Space Activities*, Office of Technology Assessment, OTA-ISC-239, US Congress, Washington, DC.

Vaughan, R.A. and Cracknell, A.P. (1994), *Remote Sening and Global Climate Change*, NATO ASI Series, Springer-Verlag, Berlin.

Chapter 2

Allan, T.D. (ed.) (1983), *Satellite Microwave Remote Sensing*, Ellis Horwood, Chichester.

Curran, P.J. (1985), *Principles of Remote Sensing*, Longman, London.

Feinberg, G. (1968), Light, *Scientific American*, **219**, 50–58.

Holz, RX. (ed.) (1973), *The Surveillant Science: Remote Sensing of the Environment*, Part 1, *The Electromagnetic Spectrum – Energy for Information Transfer*, Houghton Mifflin, Boston, MA, 1–27.

Lockwood, J.G. (1974), *World Climatology: an Environmental Approach*, Edward Arnold, London.

McAllister, L.G. and Pollard, J.R. (1969), Acoustic sounding of the lower atmosphere, in *Proceedings of the Sixth International Symposium on Remote Sensing of the Environment*, Ann Arbor, Michigan, 436–450.

Rees, W.G. (1990), *Physical Principles of Remote Sensing*, Cambridge University Press, Cambridge.

Slater, P.N. (1980), *Remote Sensing: Optics and Optical Systems*, Addison-Wesley, Reading, MA.

Tsang, L., Kong, J.A. and Shin, R.T. (1985), *Theory of Microwave Remote Sensing*, John Wiley, New York.

Chapter 3

Aracon (1971), *The Best of Nimbus*, Contract No. NAS-5-10343, Aracon, Concord, MA.

Barnett, J.J. and Walshaw, C.D. (1974), Temperature measurement from a satellite, in *Environmental Remote Sensing; Applications and Achievements*, Barrett, E.C. and Curtis, L.F. (eds), Edward Arnold, London, 185–214.

Carsey, F. and Zwally, H. (1986), Remote sensing as a research tool, in Untersteiner, N. (Ed.) *The Geophics of Sea Ice*, Plenum, New York, 1021–1098.

ESA (1987), *Remote Sensing for Advanced Land Applications*, ESA Land Applications Working Group, ESA SP-1075, European Space Agency, Paris.

Fleagle, R.G. and Businger, J.A. (1963), *An Introduction to Atmospheric Physics*, Academic Press, New York.

Laing, W. (1971), Earth resources satellites, in *A Guide to Earth Satellites*, Fishlock, D. (ed.), Macdonald, London and Elsevier, New York, 69–91.

NASA (1987), *HIRIS: High Resolution Imaging Spectrometer, Science Opportunities for the 1990s, Earth Observing System, Vol. Hc*, EOS Instrument Panel Report, National Aeronautics and Space Administration, Washington, DC.

NOAA (1985), *The Space Station Polar Platform: NOAA Systems Considerations and Requirements*, NOAA Technical Report NESDIS 22, US Department of Commerce, Washington, DC.

Peel, R.F., Curtis, L.F. and Barrett, E.C. (eds) (1977), *Remote Sensing of the Terrestrial Environment*, Butterworths, London.

Polcyn, F.C., Spansail, N.A. and Malida, W.A. (1969), How multispectral sensing can help the ecologist, in *Remote Sensing in Ecology*, Johnson, P.L. (ed.), University of Georgia Press, Athens, GA, 194–218.

Schanda, E. (ed.) (1976), *Remote Sensing for Environmental Sciences*, Springer-Verlag, Berlin.

Schmugge, T. *et al.* (1973), *Microwave Signatures of Snow and Fresh Water Ice*, Publication No. X-652-73-335, NASA, Greenbelt, Maryland.

Sellers, W.D. (1965), *Physical Climatology*, University of Chicago Press, Chicago.

Yentsch, C.M. and Yentsch, C.S. (1984), Emergence of optical instrumentation for measuring biological parameters, *Oceanographic Marine Biology Annual Review*, **22**, 55–67.

Chapter 4

Babichenko, S. and Reuter, R. (1997), *Proceedings, 3rd EARSeL Workshop on Lidar Remote Sensing of Land and Sea*, Tallinn, Estonia 17–19 July 1997, EARSeL, Paris, France.

Chen, H.S. (1985), *Space Remote Sensing Systems – an Introduction*, Academic Press, Orlando, FL.

Choudhury, B.J., Kerr, Y.H. Njoku, E.G. and Pampaloni, P. (1995), *Passive Microwave Remote Sensing of Land-Atmosphere Interactions*, VSP, Utrecht, The Netherlands.

ESA (1991), *Report of the Earth Observation User Consultation Meeting*, ESA SP-1143, European Space Agency, Paris.

Gjessing, D.T. (1978), *Remote Surveillance by Electromagnetic Waves*, Ann Arbor Science Publishers Inc., Ann Arbor, MI.

Grant, K. (1974), Side-looking radar systems and their potential application to Earth resources surveys, *ELDO/ESRO Scientific and Technical Review*, **6**, 117–136.

Hyatt, E. (1988), *Key Guide to Information Sources in Remote Sensing*, Mansell, London.

Laird, A.G. (1977), Passive infrared sensing of the environment, in *Remote Sensing of the Terrestrial Environment*, Peel, R.F., Curtis, L.F. and Barrett, E.C. (eds), Butterworths, London.

Lillesand, T.M. and Kiefer, R.W. (1994), *Remote Sensing and Image Interpretation*, 3rd Edn, Wiley, New York.

NASA (1984), *Earth Observing System, Science and Mission Requirements* (Working Group Reports Vols I and II), Technical Memorandum 86129, National Aeronautics and Space Administration, Greenbelt, MD.

NASA (1993), *EOS Reference Handbook*, NASA, Washington, DC.

NASA (1987), *High Resolution Multifrequency Microwave Radiometer (HA4MR)*, *Earth Observing System*, Vol. IIe, EOS Instrument Panel Report, National Aeronautics and Space Administration, Washington, DC.

Natural Environment Research Council (1974), *Remote Sensing Evaluation Flights, 1971*, Curtis, L.F. and Mayer, A.E.S. (eds), NERC Publication Series C, No. 12.

Naval Research Laboratory (1989), *DMSP SSM1I Calibration Validation, Final Report (Vol. 1)*, NRL, Washington, DC.

NOAA (1987), *Passive Microwave Observing from Environmental Satellites, a Status Report*, NOAA

Technical Report NESDIS 35, US Department of Commerce, Washington, DC.

Schanda, E. (ed.) (1976), *Remote Sensing for Environmental Sciences*, Springer-Verlag, Berlin.

Smith, J.T. (ed.) (1968), *Manual of Color Aerial Photography*, American Society of Photogrammetry, Falls Church, VA.

Trevett, J.W. (1986), *Imaging Radar for Resources Surveys*, Chapman and Hall, London.

Chapter 5

Austin, R.G. (ed.) (1994), *RPVs: Remotely Piloted Vehicles*, Supplementary Papers and Workshop Proceedings, Eleventh International Conference 12–14 September 1994, Bristol, UK.

Austin, R.G. (1994), The Rotary Wing RPV System, *RPVs Remotely Piloted Vehicles*, Supplementary Papers and Workshop Proceedings, Eleventh International Conference 12–14 September 1994, Bristol, UK.

ESA (1979), *Spacelab Users Manual*, ESA Scientific and Technical Publications Office, European Space Agency, Paris.

ESA (1986), *Europe from Space*, ESA SP-258, European Space Agency, Paris.

Harris, R.A. (1988), *Satellite Remote Sensing*, Routledge and Kegan Paul, London.

Hart, D. (1987), *The Encyclopaedia of Soviet Spacecraft*, Bison Books, London.

Lanzl, F. (1996), *Photogrammetry and Remote Sensing from Space*, Proceedings of the MOMS-02 Symposium, Koln (Cologne), Germany 5–7 July, 1995, EARSeL, Paris, France.

Mather, P.M. (ed.) (1995), *Terra 2: Understanding the Terrestrial Environment, Remote Sensing Data Systems and Networks*, John Wiley, Chichester.

NASA (1976), *Landsat Data Users Handbook*, Document No. 76SDS-4258, Goddard Space Flight Center, Greenbelt, MD.

NOAA (1989), *NESDIS Office of Research and Publications Research Programs*, US Department of Commerce, Washington, DC.

NRSC (1987), UK *National Remote Sensing Centre Data Users Guide*, NRSC, Farnborough.

Sabins, F.F. (1998), *Remote Sensing Principles and Interpretation*, 3rd. Edn. Freeman, San Francisco.

Velten, E. (1976), *Study on Geosynchronous Multidisciplinary Earth Observation Satellite*, Final Report, ESA Contract No. SC/127/76/HQ, Dornier/BAC/Sodetag.

Chapter 6

Becker, R, Bolle, H.J. and Rowntree, P.R. (1988), *The International Satellite Land-surface Climatology Project*, ISLSCP-Report No. 10, ISLSCP Secretariat, Free University of Berlin.

Beckett, P.H.T. (1974), The statistical assessment of resource surveys by remote sensors, in *Environmental Remote Sensing; Applications and Achievements*, Barrett, E.C. and Curtis, L.F. (eds), Edward Arnold, London, 11–27.

Curran, P.J. (1985), *Principles of Remote Sensing*, Longman, London.

Curtis, L.F. (1971), *Soils of Exmoor Forest*, Soil Survey Gt. Britain, Rothamsted Experimental Station, Harpenden.

Curtis, L.F. (1973), The application of photography to soil mapping from the air, in *Photographic Techniques in Scientific Research, Vol. 1*, Cruise, J. and Newman, A.A. (eds), 57–110.

Curtis, L.F. (1974), Remote sensing for environmental planning surveys, in *Environmental Remote Sensing, Applications and Achievements*, Barrett, E.C. and Curtis, L.F. (eds), Edward Arnold, London, 88–109.

Curtis, L.F. and Hooper, A.J. (1974), Ground truth measurement in relation to aircraft and satellite studies of agricultural land use and land classification in Britain, *Proceedings, Frascati Symposium on European Earth Resources Satellite Experiments*, European Space Research Organisation, 405–415.

Pickles, J. (1995), *Ground Truth: The Social Implications of Geographic Information Systems*, Guilford Press, New York, NY.

Milton, E.J. (1987), Principles of field spectroscopy, *International Journal of Remote Sensing*, **8(12)**, 1807–1827.

Sorensen, B.M. (1979), *The North Sea Ocean Colour Scanner Experiment*, 1977, Joint Research Centre, Ispra, Italy.

Townshend, J.R.G. (ed.) (1981), *Terrain Analysis and Remote Sensing*, George Allen and Unwin, London.

Chapter 7

American Society of Photogrammetry (1983), *Manual of Remote Sensing*, 2nd Edn, Vols I and II), AMS, Falls Church, VA.

Curtis, L.F. (1973), The application of photography to soil mapping from the air, in *Photographic*